From ants to plants, toadstools to tsunami . . . this is the book that will give you hours of pleasure as you learn about science—in fast, easy-to-follow question and answer format. Once you pick it up, you'll have trouble putting it down!

Find out about . . .
• EARTHQUAKES • VOLCANOES • MENTAL TELEPATHY • PLANTS • STARS • METEORITES • SPACE TRAVEL • DINOSAURS • GEMS • BLACK HOLES • SNAKES • INSECTS • SPIDERS • BEARS • BATS • WOLVES • CATS • PYRAMIDS • MUMMIES •
. . . and much more!

SCIENCE TRIVIA

From Anteaters to Zeppelins

FROM ANTEATERS TO ZEPPELINS

CHARLES J. CAZEAU

BERKLEY BOOKS, NEW YORK

This Berkley book contains the complete
text of the original hardcover edition. It
has been completely reset in a typeface
designed for easy reading, and was printed from
new film.

SCIENCE TRIVIA

A Berkley Book / published by arrangement with
Plenum Publishing Corporation

PRINTING HISTORY
Plenum Press edition published 1986
Berkley edition / June 1992

For information address: Plenum Press,
a Division of Plenum Publishing Corporation,
233 Spring Street, New York, New York 10013.

ISBN: 0-425-13306-0

A BERKLEY BOOK ® TM 757,375
Berkley Books are published by The Berkley Publishing Group,
200 Madison Avenue, New York, New York 10016.

PRINTED IN THE UNITED STATES OF AMERICA

10 9 8 7 6 5 4 3 2 1

To my daughters,
Sharon L. Gurri-Cazeau and Suzanne C. Cazeau,
who ask many questions

Preface

A major purpose of this book is to entertain. It may also serve to satisfy that curiosity all of us have about things both mundane and exotic. Do anteaters eat only ants? Which came first, the chicken or the egg? What are the black holes astronomers talk about? Could San Francisco fall into the sea if an earthquake occurs? Are there such creatures as vampires? Why did the ancient Egyptians mummify their dead? In exploring the answers to such questions, the reader may come to a better understanding of the workings of nature and the human mind as well.

These questions and answers are an outgrowth of a newspaper column entitled "Let's Explore" published over several years by the Gannett News Service and the *Buffalo News*. In this volume, we have arranged the questions into related categories starting with outer space and the solar system, and then coming back to the earth to look at the materials on the earth's crust and the forces and processes that affect the earth. We also consider the evolution of life on the earth, both plant and animal, including questions about their habits and behavior.

We also address man's past from primitive cavemen

of the old world to the mysterious Maya of the new world. In the process, questions arise concerning man's beliefs and the origins of demons, witches, and ghosts. Even today, many people claim to be able to foretell the future and possess other powers of the mind. What is the truth here?

On a philosophical note, there will always be unanswered questions, and even questions about our world and universe that cannot yet be framed. We would hope that whatever answers we have provided to some questions may provoke the reader into further investigation of the subject. We have been asked, "Don't you ever run out of questions?" Never. Each question spawns still others, and the expanding sea of human knowledge touches upon broadening shorelines of the unknown.

We would like to acknowledge the help of others during the preparation of this book. Sharon L. and Francisco D. Gurri-Cazeau assisted with several questions and are frequent contributors to the still continuing "Let's Explore" column. Linda Greenspan Regan, Trade Book Editor for Plenum Publishing Corporation, was an active partner in the book's production from the beginning. We acknowledge the help of Peter Strupp and Donald Arters, also of Plenum. We would also like to acknowledge our colleague, Stuart D. Scott, Jr. Finally, we thank all those individuals who asked the questions.

CHARLES J. CAZEAU

Contents

1
OUTER SPACE

Beyond the Solar System

Q *Television sets have been named after them, but I still don't know what a quasar is; do you?*

A No, and astronomers aren't sure either. They were first discovered in the 1960s and appeared to be incredibly bright objects at a distance from the earth of 10 billion light-years (more than 60 trillion miles away). They also are traveling at unbelievable velocities, such as 177,000 miles per second. Beyond those simple facts, we can only speculate. Because they are so distant, they may be objects formed early in the history of the universe, but even this idea is challenged by some astronomers who claim the red-shift, used to calculate distance and velocity, is defective. Others claim that quasars represent matter and antimatter in collision, which theoretically would result in a gigantic explosion and obliteration. Other than this, we must await further discoveries to increase our understanding of these ''quasi-stellar'' sources.

Q *What are black holes?*

A The thought that these objects might exist in outer space was first mentioned about 200 years ago by Laplace. The idea was brought to life again in 1916 and within the past 15 or 20 years, astronomers have started to get serious because our instruments for study are better. Their existence is logical. A star uses its nuclear fuel and in that process the outward pressure counterbalances the gravitational force of the mass of matter that makes up the star. When the nuclear fuel is all used up, gravity takes over and the former star collapses into a small ball of incredibly dense matter. For example, to make the earth a black hole, the substance of the earth would have to be compressed into a ball about one-half inch in diameter! The pull of gravity is so great that not even light can escape and so the black hole is invisible. We can only detect black holes by their effects on nearby stars and stellar matter. There may be many black holes, but only one that has been found so far seems to be certain and that is called Cygnus X-1.

Q *If black holes can't be seen, how do we know they are there?*

A There are many things that exist but are nonetheless invisible to us, like the air. We become aware of them by indirect means; for example, the movement of tree limbs in a breeze tells us air is real.

Black holes may represent the end point in stellar evolution—a star that has consumed all of its hydrogen fuel. There is a collapse and matter becomes incredibly packed together and the former star becomes dark. The star Sirius has a dark companion that is the size of the earth but has the same amount of matter in it as our sun. In a black hole, matter is even more compressed and has a terrific gravitational pull that can shift stars from their courses and affect radio signals. Astronomers have observed these effects around seemingly empty areas of space, and that is why we know something must be there. For the time being, we call them black holes. There may be thousands, or even millions of them, in the universe.

Q *What would happen if an astronaut tried to land on a black hole? Would he be able to get out?*

A What would happen to the astronaut is frightening. You have to understand the power of gravity. Suppose you asked a person to lift a 1000-pound weight here on the earth. They couldn't budge it an inch. Gravity is holding the block to the earth. The same force in a black hole is perhaps a million times greater. Thus, an astronaut within the gravity field of a black hole would be pulled apart. If he were headed toward the black hole feet first, the pull would be greater on the feet than the head. He would be dismembered by the force. Shortly after that, he would literally be shredded into his component atoms and molecules. You can pass by a black hole safely if you stay outside the "horizon," which is the point of no return. Within the horizon, all objects would be trapped and destroyed. Our astronauts are well aware of this should they some day rub shoulders with a black hole, but that day is probably far in the future.

Q *I have heard about regular radio signals from outer space. Could these represent communication attempts by an alien intelligence?*

A There is a very good chance they are not. Outer space is full of a variety of radio noise from beyond our galaxy as well as within it. Some signals do seem to have a pattern, but there are many regular or periodic phenomena in nature not associated with intelligence. Even many stars pulse at periodic intervals.

Here on the earth, Old Faithful geyser erupts rather regularly, and without any intelligent beings working the valves. Despite our negative response to your question, we do not dismiss the possibility of life elsewhere in the universe. However, planetary systems with life would be so incredibly far apart that communication, even by radio, would be a tall order.

Q *Why are scientists deliberately using radio waves in an attempt to contact aliens? This could be dangerous.*

A Perhaps, but we do not think such attempts to be dangerous, for two reasons. One, the communication, if it comes, will be at long distance like a phone call. Even if our galaxy is teeming with other intelligent life (as many scientists believe), interstellar distances are so incredibly vast that even with travel at the speed of light, it would take many years to establish physical contact with extraterrestrial life. It is not like dropping over to your neighbor's for a cup of coffee. Second, for the sake of drama, Hollywood and popular writers have conveyed the impression that all aliens are filled with menace and are out to get us in some way. There is no basis for this. In our view, any life form capable of achieving space travel at the interstellar level will not have barbarian reflexes or urges to kill or barbecue everybody on sight. We would expect alien contact, if it comes, to more likely be one of peace and exchange of mutually beneficial knowledge. At least, let us hope so.

Q *Is the idea of antimatter just an invention of science fiction?*

A It is by no means a science fiction gadget hatched as a plot for a *Star Trek* episode. Nor is the idea new. Long before World War II, the concept was accepted on theoretical grounds. In 1932, the positron was discovered. This is the opposite of an electron, having a positive rather than a negative charge. Later, in 1955, the antiproton was discovered, carrying a negative charge, opposed to our "ordinary" proton, which carries a positive charge. It is known that if an atom of antimatter were to collide with an atom of our matter, a spontaneous burst of radiant energy would occur, dwarfing the energy released during an ordinary nuclear blast. Some scientists speculate that somewhere in the universe there are galaxies of antimatter. None has as yet been found. Still, a real possibility is that they exist. They would not look any different than constructions of matter in our own familiar universe. Yet if they came together with matter, total annihilation would result. Let us hope that we do not meet up with antimatter.

Q *If the stars we see at night are moving as fast as astronomers have reported, why are they always in the same place?*

A They are moving, of course, and at breathtaking speeds. Some of the more distance stars are racing through space at many thousands of miles per second. Their apparent lack of movement is due to the vast distances that separate us. Apparent motion is a function of distance and the relationship between the observer and the moving object. Anything moving away from an observer lacks any great motion. The stars are moving away from us. An example from our common experience is a jet airplane taking off and moving away from us. The farther away it gets, the less movement you see. Indeed, in some cases, the aircraft seems to be standing still in the sky. Yet we know the plane is traveling at speeds of about 600 miles per hour. Thus, stars, many billions of miles away, also appear motionless. But given enough time, they will shift position. The present North Star has not always been the polestar, and over thousands of years, all stars will be in new positions.

Q *I've read that if you were to depart on a space voyage at the speed of light and return to earth after 1000 years' earth time, you would have aged only 50–60 years. Could you explain this in layman's language?*

A This concept is part of Einstein's theory of relativity, which has revolutionized our ideas about space and time. In essence, Einstein says that the faster you move, especially as you approach the speed of light, the more time slows down relative to a stationary observer. This applies to biological processes as well as mechanical objects such as clocks. Thus, a man aboard a spaceship, moving close to the speed of light, will experience a slowing of heart beat and respiration, but to him everything will seem as usual. If he sleeps for eight hours, that would amount to several days' duration to an earthly observer.

The flow of time is not fixed and unchanging as we all thought because we are so accustomed to ordinary earth rates of travel. Einstein was proved correct as early as 1936 at Bell Laboratories when radioactive hydrogen atoms were accelerated to high velocities and compared to hydrogen atoms at rest. The frequency of vibration (a sort of clock) of the moving atoms was reduced as Einstein predicted. When asked, Einstein said, "Is it any more strange to assume that moving clocks slow down than to assume that they don't?" Whenever Einstein's theories have been put to the test, they have been shown to be correct.

Q *Are planets rare freaks of nature? If there aren't a great many of them, the possibility of life existing elsewhere in the universe would be small.*

A Your point is well taken. How common are planets? No one knows for sure because at interstellar distances, they are too tiny to be seen using telescopes. Yet we are here and our sun does have a family of planets. Are we a freak? Years ago, scientists were inclined to believe so. It was held that our sun came into a near collision with a passing star. This was conjectured to have caused gigantic gaseous eruptions from the sun, eruptions that fell into orbit about the sun and cooled to form the planets. An accident, in other words. This idea is no longer held. It is now believed that planetary systems are a normal part of stellar evolution, and if that is the case, then there are billions of planets circling stars, and the possibility of other life is magnified. Even if we take the dim view that only one set of planets is produced per galaxy by accidental collision, that gives ten billion planetary systems. If only one out of 1000 supported life, that would be ten million planets with life. If we are only average, then there might be five million civilizations with life higher than our own.

Q *Is there any possibility that life like ours could form on a planet cold enough that all water is present as ice?*

A Scientists have thought about this. We regard a life form as an energy system that requires some kind of liquid to carry out the chemical reactions that make life possible. Here on the earth, life is essentially protein-in-water. Water is the vital liquid.

On a planet where water is always in a solid form (ice), there could not be earth-type life. However, bear in mind that other substances such as ammonia and methane, which are gases on the earth but would be liquids on a cold planet, could be the basis for some kind of life on a cold planet. With ammonia, this could be an ammonia-in-protein kind of life at –30°F; with methane, a form of life involving fatty substances in methane at 300°F. We can only guess what form such life would take, and whether or not it would be intelligent. One thing is for sure: such life would think our earth to be intolerably hot! We may never know whether or not such life exists.

Q *Could life occur on planets so hot that there are gases but no liquids, which you say are necessary for life?*

A We would have to say that some hot planets, like cold planets, could sustain some form of life. At temperatures between 235 and 800°F, sulfur is a liquid, and at even higher temperatures, silicon is a liquid. Both are solid at normal earth temperatures. Yet both are "joiners" like carbon, the basis for earth life. A joiner is an element that can enter into an astounding number of combinations with other elements to form a variety of substances, including substances allowing the possibility of life. What this means is that we have to entertain the idea that on some remote, scorched planet, there may be life forms, even intelligent, that might look at our world and declare it too cold to harbor life "as they know it!"

Q *In previous questions, you implied that there's life on other planets. Isn't this belief based entirely on probabilities, a sort of numbers game?*

A Yes and no. If we are talking only about our kind of life, that life is based on the elements carbon, hydrogen, oxygen, nitrogen, and a little phosphorus and sulfur combining to form such vital things as proteins, water, fat, and carbohydrates needed for life on the earth whether it be human or cockroach. Analysis of light received from distant stars has shown that these necessary elements are present throughout the universe. At this point, the numbers game comes in. If there were only a few planets, the chances of them being earthlike and possessing these important elements would be pretty slim. But the trillions of stars that we know exist suggest that there must also be at least billions of planets of varying compositions. As the number of these planets increases, so too does the possibility that there are planets by the millions with the right ingredients for life to start, given enough time. Life exists here on the earth. Should we be so unique?

Q *If an alien were to land on earth, what would he (or she) probably look like?*

A We believe extraterrestrial life, if it exists, could be of almost any form imaginable. Think of the earth's diversity of life, such as the whale, maple tree, and ant.

Even more bizarre life forms might arise on a planet whose environment differs from that of the earth.

If we are talking about mobile, intelligent life (we haven't seen any maple trees piloting space ships lately), then some scientists argue that the humanoid form has many advantages. Our major sense organs are "clustered" in one package (the head) elevated above the ground. Evolutionary development might favor this arrangement in coping with environmental situations on other planets as well as our own. Hands with opposable thumbs provide the capacity for toolmaking, another big advantage. Of course, these are speculations, and until "contact" is made, it must remain another area of the unknown.

Q *If aliens from another planet were to land in Washington, what would happen?*

A This is a speculative question, but an interesting one. It would be logical to assume that these beings are intelligent and have technological skills. However, these traits in no way guarantee benevolence. The Nazis are a reminder. But for such an intelligent and advanced life form to seek us out and try to establish contact with a cultural group of which they know so little would lean in favor of a friendly encounter. To reach the earth bespeaks a desire to acquire knowledge, and perhaps a desire to impart knowledge, which would be all to the good. We often think of space "invaders" as being hostile to us, but what about the other side of the coin? Aliens would be arriving at an earth on which humans are engaged, at this moment, in numerous wars around the globe. Would this seem rational and desirable for an alien civilization? Probably not. If our astronauts reached an inhabited planet where the same thing was occurring, we might not want to stay either. In any event, such an encounter is not likely to occur in the foreseeable future, given the vastness of outer space.

Q *What is the difference between astrology and astronomy?*

A Both have the same roots. Astrology came before astronomy, perhaps 4000 to 5000 years ago in Mesopotamia, where it was believed that the stars and planets, in their movements and positions, had an effect on people. Observers of the heavens kept very good records of phases of the moon and the rising and setting of individual planets, so that astrology can be rightly called the earliest exact science. Beginning in the Middle Ages, some observers came to realize that the stars did not control human events, and split off from those who did. Thus, astronomy was born. Today, astronomy deals with the exploration of the universe and how it evolved and is evolving. Astrology continues to be what it was 5000 years ago, a pseudoscience claiming to give insight about an individual's destiny based on the position of stars and planets, for which there is scant scientific support.

Within the Solar System

Q *We are constantly bombarded by cosmic rays. Just what are cosmic rays and are they dangerous?*

A Cosmic rays consist of high-speed particles hitting the earth. One source of these is the sun, but most of them come in from outside the solar system, traveling near the speed of light. Two types of particles are known; primary and secondary. The primary particles are protons that reach the earth from outer space. As they enter the earth's atmosphere, they collide with atoms, imparting speed to them, something like hitting one billiard ball with another. These are the secondary particles. Since we all have been bombarded by these particles throughout our lives, they are not dangerous to us even though they are a form of radiation. The study of cosmic rays, discovered in 1903, continues and can be considered a branch of astronomy.

Q *What are sunspots and what effect do they have on humans, if any?*

A Sunspots appear as dark specks on the solar surface and were first observed by Galileo. They seem to be large "whirlpools" of cooler solar matter, and are associated with a magnetic field.

Sunspot activity occurs in 11-year cycles. High levels

of sunspot activity cause radio interference. There is some suggestive evidence that sunspot activity affects our weather, and if it does, then life, too, is affected. Thicker growth rings in trees can occur every 11 years, corresponding to the sunspot cycle. Direct effects on human activity are not exactly known.

Q *Are the northern lights caused by burning gases in the atmosphere?*

A Not exactly, but the cause of the northern lights (or the aurora borealis) is not well understood. We know that they occur most frequently after there have been flares of other activity on the sun, and it seems the hydrogen from the sun is reaching the earth and interacting with the earth's magnetic field to produce the display of "curtains" and other phenomena best seen in northern latitudes. It is amazing that 100 million hydrogen particles strike one square inch of the atmosphere per second during an auroral display. This is what causes the light, which is more of an electrical charge than a burning gas. On rare occasions, these lights are seen in southern Europe and in southern states. Years ago, to the people living there, such lights were held by superstition to be a warning of some kind of disaster.

Q *What is the earliest reference to eclipses? They must have been frightening to primitive tribes.*

A Yes, they were frightening and even today they still scare people throughout the world. A solar eclipse interrupts the normal rising and setting of the sun routine and is viewed as a calamity. Early peoples beat drums and shot arrows into the air to drive away the "monster" that was eating up the sun. The fact that a total eclipse lasts only eight minutes led them to think that their demonstrations were effective. Eclipses have been recorded and even predicted for a long time. The earliest reference we have is from Chinese astronomers who noted an eclipse on October 22, 2136 B.C. An eclipse of the moon took place on April 2, 1492. At that time Christopher Columbus was attempting to obtain provisions from natives of the island of Jamaica, without success. Knowing of the coming eclipse, Columbus warned the natives he would do something to the moon unless they gave him food. After the eclipse, Columbus got what he asked for from the frightened natives.

Q *A friend showed me a small black stone and told me it was from outer space. Can he be right?*

A Of course we can't know if your friend is an expert on the subject, but there are natural glass objects known to geologists as tektites, and at least some of them are thought to have passed through the earth's atmosphere. Most of them are small, ranging from pebble size up to as big as your fist, and are black. Some are green or yellow. The surface of these objects shows signs of melting (in fact, the word "tektite" is derived from a Greek word that means melted). Some scientists believe that although tektites result from meteorite impact, they did not in fact come from outer space. The basis for this idea is that many tektites are related chemically to the surrounding rock and soil in which they are found. Tektites may be as old as 50 million years. Others are much younger. Astronauts have found similar objects in lunar soils.

Q *Could life on the earth have started from a meteorite landing here? There is carbon in meteorites.*

A It would be an extremely long shot (no pun intended). Yes, there is carbon in some meteorites (the iron-nickel variety) but it takes more than the mere presence of carbon for life to form. Laboratory experiments suggest that hydrogen and oxygen are also necessary as well as the right environmental circumstances for amino acids and thus protein to form. And it would take a long time. Another point is that there was already plenty of carbon available on the original earth without depending on carbon-bearing meteorites. Of perhaps greater significance is that the testimony of meteorites indicates that carbon is present throughout the solar system and probably the universe so that carbon-based life (such as ours) may exist elsewhere. The idea that life on the earth could have come about from a source in outer space is known as the cosmozoic theory, and is held in little regard by scientists today. All of the geological and biological evidence suggests that life originated on our planet. Science has not yet been able to create life although there have been provocative steps in this direction. The origin of life is still a mystery.

Q *Concerning the idea that life may have originated from meteors, where do these meteors come from?*

A We would clarify that we do not believe that meteors brought life to the earth even though they contain, in some cases, the element carbon, which is the basis for life on the earth. Meteors seem to be the leftover debris resulting from the creation of the solar system. Some may have resulted from the disintegration of larger asteroids. There are two major categories of meteorites (i.e., meteors that have reached the earth): the stony meteorites and the iron–nickel meteorites. These contain all the elements that we find here on the earth but in some instances, meteorites contain exotic minerals not found occurring naturally on the earth.

Meteoric material is abundant in space and the earth receives many tons of it every day. Most of it falls as "dust" after breaking up in the earth's atmosphere. Occasionally, as we all know, larger meteors strike the earth, such as the big one at Meteor Crater, Arizona. Meteor showers also take place. Back in 1868, it was estimated that 100,000 meteors fell in a shower over Poland. Most of these pieces were no larger than grains of rice, however, the larger meteors seen in the sky are bright with fiery tails. If they land, they are still hot, but this heat is superficial. After a while they may be covered with frost. This is because they still retain the cold of outer space inside them.

Q *Is it true that meteorites that crash on the earth come from the moon?*

A We would say that only some do come from the moon. This is because in the remote past the moon itself was hit by large meteors or small asteroids so that fragments were catapulted into space. The mare areas—the dark spots—on the moon represent these sites of impact. Eventually, some of these lunar fragments reached the earth. The reason we know they are from the moon is we have moon rocks to compare them with. It is amusing that we went to the moon to bring home moon rocks when there already were moon rocks on the earth, but we couldn't identify them until we had gone to the moon for a sample of its rocks. The moon rocks that arrived here as meteorites came from the lunar "highlands" (the bright areas) and are breccias, a rock consisting of angular fragments welded together.

Q *What are the big dark spots that make up the eyes of "the man in the moon"?*

A Early in the moon's history, it was bombarded by big asteroids that struck the surface with such force that gaps were created reaching into the moon's interior. At that time, the inside of the moon was still molten with lava material. This lava upwelled to the surface and spread out as circular "lakes" of the dark lava known as basalt. It cooled and solidified to form what are called the mare areas of the moon. "Mare" means oceans or seas, as the early astronomers thought them to be. Our astronauts have landed in these areas and brought back to the earth samples of this material. These lunar rocks of lava material have been found to be little different from the black lava that today erupts from volcanoes in Hawaii. A clear difference is that the lava on the moon is as fresh as when it erupted because there is little chemical weathering on the moon—owing to the absence there of air and water—while on the earth, lava and other rocks undergo much chemical change due to weathering.

Q *Is there any truth to the theory that the moon is a captured satellite of the earth? I am told this created disruptions that caused the continents to move around.*

A There is no scientific evidence that this is so, although it is an interesting idea. We know from other lines of geological evidence that the continents have been moving around since very early in the earth's history due to other causes than the moon. The moon appears to have formed at the same time as the earth, according to analysis of rock samples brought back from the moon by astronauts. A striking piece of evidence that the moon has always been where it is comes from a study of ancient corals that lived in the sea more than 400 million years ago. Microscope viewing of these old corals shows growth layers that reflect a tidal influence, something that could not occur without a moon up there causing tides. So, at least the moon was in place as long ago as 400 million years, and most geologists believe the moon was formed at the same time as the earth.

Q *How big is the largest asteroid and what would happen if it hit the earth?*

A The largest known asteroid is called Ceres, lying in an orbit between Mars and Jupiter along with thousands of other asteroids. It is about 500 miles in diameter. The possibility of such an asteroid colliding with the earth is extremely remote. However, should that occur, it would be a disaster of unparalleled magnitude. Were it to smack into the central United States, we could imagine it creating a crater perhaps 1500 miles in diameter and an upwelling of molten material from within the earth. It is doubtful if humanity could survive such a cataclysm, but it is something not to worry about. On a smaller scale, some meteors originating in the asteroid belt have struck the earth and provide scientists with useful information about the age and composition of astral bodies.

Q *Halley's comet came in 1986. If it had struck the earth, would civilization have been destroyed?*

A First of all, the chances of Halley's comet striking the earth were infinitesimal. It lies along a very predictable orbit that allows no likelihood of a collision. There are asteroids that come very near to the earth, even crossing its orbit, but the chances of them hitting us are exceedingly small, although greater than for Halley's comet.

The matter that makes up a comet is quite small. As our colleague L. Sprague de Camp has observed, if you could take all the matter in the tail of an average comet and compress it to the density of iron, you could put it in your suitcase! The greatest devastation we can imagine if Halley's comet struck the earth in a populated site is a loss of life equal to that resulting from automobile accidents in a single month. It should be noted that small asteroids and other space matter do reach the earth daily, but they mostly burn up in the atmosphere before reaching the ground. Occasionally, but at long time intervals, they crash, leaving traces such as the Barringer crater in Arizona.

Q *The craters on the moon are perfectly circular. How could the meteors that made them always descend vertically to the moon's surface?*

A You are assuming that circular craters can only be made by vertical impact. Laboratory experiments have been carried out using gas guns and simulated lunar soil to see how these craters might have formed. Results show that if a meteor crashes at any angle greater than 15 degrees, a circular crater is always formed.

There are elongated craters on the moon, or "skip marks" as they are called, produced by meteors slanting in to the lunar surface at less than 15 degrees. Some of the craters on the moon have been formed by volcanic action as well.

Q *A big explosion in Siberia in 1908 was supposed to be that of a meteorite. I hear now that it was really a giant, nuclear-powered spaceship because the area of the blast was radioactive. Is this true?*

A The Tunguska explosion was certainly one of the biggest in history. It flattened about 45,000 trees. Whatever the object was, it didn't strike the earth because there is no crater. It exploded four to six miles above the earth. Some people who saw it are still alive. It could have been a meteorite or a large comet. It has also been suggested that it was a "black hole" or perhaps a big chunk of antimatter. The weight of evidence favors a meteorite. We have heard the story that it might have been a spaceship. However, the site of the explosion was not particularly radioactive, and so the spaceship idea is only an interesting speculation. We will perhaps never know what it was. If the explosion had taken place over a major city, it could have been the worst calamity in human history.

Q *I know that Galileo discovered the moons of Jupiter with a telescope. Did he invent the telescope?*

A No, but he was the first to dramatically improve it and to achieve major discoveries using it. In 1608, an eyeglass maker named Hans Lippershey, a Dutchman, accidentally held up two large lenses while looking out a window and was startled to be able to see distant objects as though they were nearby. He assembled the lens in a tube and thus the first telescope was made; but it was only 3 power. Galileo heard about this when he was in Venice and soon made his own telescope of 32 power. He then discovered not only the moons of Jupiter but spots on the sun, the millions of stars in the Milky Way, and the phases of Venus (like the moon) that proved that all the planets moved around the sun rather than all of them circling the earth. Proclamation of that fact got Galileo into trouble with the church, and he spent the last eight years of his life under house arrest for heresy.

Q *Can it really be that Galileo was arrested and imprisoned for saying that the earth goes around the sun?*

A Yes. Back in the 17th century it was risky being a scientist because much of what they discovered ran contrary to the thinking of ecclesiastical authorities who had the power of life and death. Galileo was a genius, being a pioneer in mathematics, mechanics, and astronomy, and was widely respected during his own lifetime. Had he not been a close personal friend of the Pope, it is possible that he might have been burned at the stake. In his trial, he was forced to deny his statements that planets circled the sun and the Copernican theory was denounced. As he was at that time an elderly man, his punishment was confinement under house arrest at his villa near Florence for the remaining (eight) years of his life. While such a thing could not happen today, we still see much anti-science in attempts to ban evolution from biology textbooks. As one scientist put it, ''Galileo may be dead, but his persecutors are alive and well.''

Q *Since our space probes have reached Mars, is there an explanation as to why Mars is red?*

A Astronomers have long noted the red color of Mars as did ancient peoples who linked the planet to the god of war. The Viking landers and the Mariner orbiter, unfortunately, have not given us an unequivocal answer. It is reasonable to assume that soils, rocks, and minerals on Mars will be like or similar to those on the earth in many but not all respects. On our planet, minerals such as feldspar, hematite, and others are pinkish or reddish. Widespread occurrence of such minerals at the surface would give Mars its coloration. Present thinking is that iron oxide is the main source of the color, but there is not much oxygen on Mars so perhaps the iron oxide is merely a coating on other grains. This also occurs on the earth. Recently, it has been suggested by two Princeton University geologists that the red is due to still another iron mineral called nontronite. We will have to wait and see.

Q *Is it possible that the two satellites of Mars are artificial ones? Also, are the canals on Mars real?*

A If the satellites of Mars were artificial, it would suggest a former advanced civilization there. Our spacecraft in the Mariner program photographed these satellites in some detail. They are not artificial. The larger one, known as Phobos, is 13 miles in diameter and cratered. In shape, it was described as looking like an "Idaho potato." Deimos, the smaller satellite, is even more cratered and still too large to be an artificial satellite. In the Mariner program, 100% of the Martian surface was photographed in considerable detail and there are no canals. There are certain linear features, quite natural, that may have been mistaken by early astronomers for channels or canals. The romantic notion that a Martian civilization built the canals to draw water from the polar regions is simply not tenable. While there may be or may have been life on Mars, there is no evidence of this as yet. If there is life on Mars, it is probably a low form such as bacteria.

Q *Is it possible that the asteroid belt between Mars and Jupiter represents an exploded planet?*

A This is just what astronomers thought when they first started discovering the fragments that lie, by the thousands, in the asteroid belt. As long ago as the 18th century, the astronomers Titius and Bode had recognized that the distance between the orbits of any given pair of planets was about double the distance between the orbits of the preceding pair. This placed a planet between Mars and Jupiter and also aided in the discovery of outer planets such as Uranus. In 1801, Ceres was discovered in the asteroid belt and it was thought to be the predicted planet. However, it was rather small, only about 470 miles in diameter. Later, other small planetoids such as Juno, Pallas, and Vesta were found. The latter can sometimes be seen with the naked eye. This led to the idea that these were remains of a planet that had somehow blown up early in the history of the solar system. It is still a debated question, in which some authorities believe that the asteroid belt represents the fragments of a planet that never coagulated. Regardless of which theory is correct, it is doubtful that the asteroid belt represents the remains of an inhabited planet blown up by atomic catastrophe, as some science fiction writers have imagined.

Q *Every fourth year is a leap year. What good are they?*

A It keeps our calendar correct and the seasons of the year where they ought to be. A year is determined by the revolution of the earth around the sun, which takes 365¼ days; days are determined by the earth spinning on its axis. Each of these motions acts independently. Every four years we add a day to make things even. In other words, if the earth completed its trip around the sun in exactly 366 days, we would not need a leap year. However, there is another fine tuning that must be done, and that is to deduct a day once a century. That deduction will next take place around the year 2000. If we didn't do this, over several centuries, summer would occur in winter months and vice versa. It would be very annoying.

Q *Immanuel Velikovsky is now dead. What is the present interpretation of his famous theory that Venus nearly collided with the earth 3500 years ago and changed human history?*

A Velikovsky's book on this subject (*Worlds in Collision*) became a red-hot bestseller during the 1950s. He claimed that the near collision explained the parting of the Red Sea and other phenomena when Moses came out of Egypt. Astronomical and historical data suggest that Venus has been in its present orbit for many millions of years and it at no time wandered about the solar system in near collision with other planets.

Velikovsky drew from historical texts that mentioned

floods, fire from the sky, and earthquakes. They probably did take place in the Mediterranean area, but were most likely the effects of the catastrophic volcanic eruption of the island of Thera. The dates coincide. Had Velikovsky been aware of Thera, which at the time he wrote his books was not known, he might very well have abandoned his theory. No scientists whom we know about lend any credence to Velikovsky's ideas. However, we do admire Velikovsky for his tireless efforts and imaginative thought. Science needs men like him.

Q *Concerning the outer space probes that have been launched: When will they reach the outer planets, and what will they tell us?*

A Some of them have already reached their destinations. Pioneer probes have given us excellent pictures and data about Jupiter. In November 1980, Voyager I was sending back data about Saturn and its famous rings. Voyager II arrived in the same vicinity about ten months later. In 1987, Pioneer II will be the first man-made object to leave the solar system and head out into the universe beyond. The information gathered will tell us a great deal about the nature of the giant outer planets, and perhaps about how the solar system originated. This, in turn, will shed light on how our own earth came into existence.

Q *Our space shuttle program is terribly expensive. Couldn't the money be better spent on earth?*

A You have to look beyond the great technological achievement (which it is) to see what it means. It is actually a cheaper way to explore and utilize outer space because the shuttle is resusable. The technology developed during the program will result in spin-offs of practical applications here on the earth to benefit everybody. We need to go into outer space in order to explore the earth. The shuttle will be able to pick up and deposit orbiting satellites with specialized instrumentation for locating mineral deposits, tracking weather patterns and storms, assessing agriculture and vegetation, keeping an eye on air pollution trends, and even picking out the best fishing grounds in the world's oceans. There will be many experiments in space physics and it will be a boon to astronomy because the sun, planets, and stars can be studied without the obscuring atmosphere. Eventually, larger, live-in satellites will be assembled in orbit, piece by piece. Perhaps some day, the shuttle will be used to assemble spaceships in orbit that will then be sent on interstellar missions.

Q *The outermost planet, Pluto, has a satellite. How can such a small object be seen from so far away?*

A In 1978, astronomer James Christy was studying photographs of the part of the sky that contains Pluto and noticed that, under great magnification, the planet seemed to have a slight bulge when it should have been circular. Looking at earlier photographs, he found that the position of the bulge changed and concluded that it must be a large satellite circling Pluto in an orbit fairly close to it. He named this satellite Charon. It was a remarkable discovery, and allowed, with the aid of sophisticated instruments, more data than we ever had before. We now know that Pluto is about 2500 miles in diameter and the big satellite, 1250. We also know that at least part of the surface of Pluto is made of frozen methane. More should be learned about the planet's composition in the near future. By the way, Pluto is not the outermost planet at the moment. It has crossed inside Neptune's orbit and, at least for a while, Neptune is the outermost planet. Even that may change because it is thought that a tenth planet lies still farther out and remains to be discovered.

Q *There was a time during the 1980s when all of the planets were aligned, and great events were supposed to have taken place. When did this happen?*

A There was such a planetary alignment in the summer of 1982 when Mars, Venus, Earth, Jupiter, Saturn, and the sun were so situated. Astrologers place great significance on such events, particularly to those born during these times. Such astronomical events have occurred several times in the past without any world-shaking catastrophes, and the people born during such times were not particularly distinguished from those born at other times. We do not visualize anything unusual happening on the earth at such times other than maybe a dog tipping over an ash can.

Q *If the earth is traveling through space at 66,700 miles per hour, why don't we feel this motion?*

A Because we are also traveling at 66,700 miles per hour. This translates to about 18 miles per second, but it is a smooth ride. It is similar to traveling at 600 miles per hour on an airplane and feeling little sense of motion. A fly buzzing around in such an airplane has even less sense that it is traveling at such speeds. It is a bit more boggling to realize that this is not the only motion we earthlings are subject to. After all, the earth is spinning on its axis at more than 1000 miles per hour, and the earth is affected by the sun's movements through space, and the sun is also involved in the general galactic rotation while the entire galaxy is hurtling outward as a part of the expansion of the universe. It is enough to make us dizzy, yet we are not.

Q *Are there people who believe we never got to the moon, that it was a giant Hollywood hoax for political reasons?*

A We have also heard this fairy tale. True, Hollywood has the talent to make many fictional things appear real, but they had no hand in this. In order to create a hoax, it would have to be restricted to only a few people. To say our voyages to the moon were a hoax is to say that thousands of individuals participated in this hoax and didn't let the cat out of the bag. Such a conclusion is groundless. But if you want solid proof other than viewing the launchings and activities on the moon via television, astronauts did return with moon rocks, which were distributed to scientists all over the world for study. Discounting that all these scientists would have also been in on the hoax, it is a simple fact that the moon rocks are different in a very important respect from similar rocks on the earth: they show virtually no weathering. They are nearly as fresh as on the day they were formed. Rocks on the earth are continually attacked by chemical weathering and this can easily be observed under the microscope. Moon rocks, in contrast, show little if any chemical weathering.

Q *Does science have any idea when the world will end, and how?*

A At the end of each century, people in white robes gather on hilltops awaiting the end of the world and it hasn't happened yet. No doubt similar groups will assume the end of the world will occur in the year 2000. There is no scientific basis for such beliefs. If you are referring to the end of our planet rather than the end of the human species, then science does have a possible answer. The origin of the earth was tied closely to the origin of the sun. The sun furnishes our heat and light. If the sun were to suddenly disappear, the earth would not last long, with temperatures steadily declining until it became a frozen chunk of rock. However, there is a more likely scenario. As the sun ages and nears the end of its existence, it will enter a nova stage and emit so much heat as to burn the earth to a crisp, with oceans boiling. Not to worry though, because the sun has several billion years more life before that happens.

2
THE EARTH

Materials and Life

Q *When did the earth begin, and how?*

A The earth came into existence about 4½ billion years ago based on several lines of evidence involving radiometric dating of rocks and the earth's crust, samples of material from the moon, and meteorites. The origin of the earth was the same as for the rest of the solar system. A large cloud of hydrogen gas slowly condensed with the center of the cloud forming the sun, while small eddies outward condensed into the planets. The earth probably passed through at least a partly molten stage as we know that the earth's interior resembles the structure of an onion, that is, concentric layers. The central layers are of heavier material than the outer layers. We know this through study of vibratory waves (seismic waves) produced by major earthquakes that travel through the earth.

Q *Life began in the sea. How did it get out on the land?*

A For millions of years, there have been tides along the world's shorelines where the beach zone has been alternately under water and then exposed to the air. It is likely that some sea creatures, marooned in this tidal zone, managed to survive until the next tide came in, and so adjusted to this mixed environment. Eventually, the fittest and hardiest of these were able to survive being out of the water for long periods. They gave rise to amphibians, and subsequently reptiles and mammals. It is interesting that we still carry part of the sea with us, because our blood resembles sea water in several respects. We might speculate on the direction of evolution had there been no moon to cause the tides. Would all life still be in the sea and the land barren? And would this aquatic life be as intelligent as man?

Q *What caused the extinction of the dinosaurs?*

A Many theories have been offered but we really don't know yet. Disease, the small size of the brain, mammals eating their eggs—all these have serious weaknesses. Any sound theory must take into account the extinction of these reptiles not only on land but also in the sea. We note that wholesale extinctions seem to take place at the same time as major changes in sea level. These changes may have affected the food supply and the climate adversely. The extinction of a species is not necessarily an unusual or preventable event. In the earth's history, three out of four of all the major groups are now extinct. Man also may well go the way of the dinosaurs, but we hope it won't be next week.

Q *Why haven't species other than the dinosaurs become extinct?*

A They have. The dinosaurs are not unique in this respect. The extinction of a species seems to be a very common result. We know from the fossil record that perhaps as many as three-fourths of all species that have arisen are now extinct. One reason for this is that as a group evolves, it tends to become more and more specialized in its life-style. This means it fits into a narrowing ecological niche, but fits extremely well into that niche. The deer becomes expert at running, the tiger well equipped for hunting. Now if there is a sudden change in the environment so that these skills are not usable, then the species dies off. Some people regard mankind as overspecialized, as in brain power. Actually, man is a more general type of life that obviously has adapted to many environments. Overspecialization would not cause us to become extinct. More likely, if we do not survive, it would be due to our inability to handle maturely the high and dangerous technology we possess.

Q *Big land dinosaurs have been found as fossils on more than one continent. How did they get across the oceans, since they couldn't swim?*

A There are several explanations. We assume that a particular kind of land dinosaur would originate in one place rather than appearing as the same species in several places on the earth. So we are most likely dealing with dispersion from one spot. The continents have shifted around and migration may have taken place when the continents abutted each other. Also, a lowering of sea level would create land bridges for animals to cross. This is doubtless how man came into North America from Asia across the Bering strait during the Ice Age. There is also the factor of time and probability for a "rare event" to happen. Given millions of years to work with, sooner or later, a rare event, such as a dinosaur being carried to another continent by a log in a storm, might occur.

It is easier to explain the spread of plant life, whose seeds can be dispersed long distances by the wind or carried by birds. This is how plant life came to the new volcanic island of Surtsey in the North Atlantic within a couple of years of it emerging from eruptions on the seafloor.

Q *Can you clarify the theory explaining the extinction of the dinosaurs as being caused by iridium?*

A It is a rather intriguing new theory that explains not only the extinction of the dinosaurs, but also every mass extinction that has occurred. It has been found that during periods of extinctions, the rocks of that time contained a higher than normal content of the element iridium. This is a metallic element resembling platinum, and is rare on the earth but fairly abundant in asteroids. It pointed toward an asteroid hitting the earth and clouding up the atmosphere with iridium-laden dust, shutting out the sun, and interrupting life processes to the point of extinction. Of great interest is that these extinctions occur about every 26 or 28 million years, leading to the idea that our sun is not alone but has a mutually orbiting stellar companion that returns to our vicinity every 26 or 28 million years, disturbs asteroid zones, and sends them flying toward the earth. Astronomers have begun searching for this other sun. Not to worry, because it is not due back for another 15 million years.

Q *What is your opinion on the recent reports of dinosaurs in Africa?*

A We have seen the tabloid headlines that make startling claims of "Dinosaurs Alive in Africa." The story behind the headline is, as usual, more realistic, since no dinosaur has in fact been found. The search is described as centered in an unexplored region of Zaire in west Africa, the former Belgian Congo; evidence, like that for Nessie (the Loch Ness monster), has been based on persistent reports of sightings by local people. They describe an animal up to 30 feet long having feet with three claws. Our guess is they will not find a dinosaur. Although new species of plants and animals are discovered often, they are of very small size and many scientists believe that no large spectacular land animals are left to be found. At least none has been found during the present century. That a dinosaur could be alive today is an exciting but unlikely idea.

Q *In school I was taught that giant prehistoric animals became extinct before humans appeared on the scene. Don't some new discoveries disprove this?*

A First, we have to know what animals you mean. During the late part of the Ice Age, such large animals as mammoths and mastodons were hunted by early man. If, on the other hand, you are thinking of films with cavemen and dinosaurs chasing each other, such dramatic scenes are only movie magic. Man's appearance came long after the dinosaurs became extinct.

But then again you may have heard of the recent important scientific discoveries in South America. Earlier it was believed that certain prehistoric mammals such as giant ground sloths and glyptodonts had disappeared before humans migrated to South America. In Bolivia, scientists have found human remains together with bones of a giant sloth dated at 7000 years ago. This means that some extinct large animals survived longer than we thought and apparently lived side-by-side with humans.

Q *On a recent fossil hunting trip, I was lucky enough to find a trilobite. Can you tell me something about these interesting fossils?*

A They first appear in the fossil record about 600 million years ago, especially in rock layers made of limestone and shale. These rocks would be from the Cambrian period. Trilobites lived on the muddy seafloor, crawling around. As they grew, they molted or cast off the hard outer skeleton just like other arthropods do today. Although they were mostly about an inch long, in later geological periods they tended to get much larger, some being two feet in length. They became extinct at the end of the Paleozoic era but have to be regarded as successful creatures, having lived more than 400 million years and producing at least 10,000 species. If you find a trilobite, you know that the rocks in which it was found are of Paleozoic age and thus the trilobite is called an index fossil.

Q *There are slabby rocks in my backyard that have butterfly fossils in them. How did they get there?*

A Butterflies and other flying creatures are extremely rare in the fossil record, and for that reason command a very fancy price. Geologists have established that almost all butterfly and moth fossils found in rocks are no older than 60 million years. The rocks in your area (we checked a geological map) suggest that your rocks are considerably older than this. What we think you've got are fossils of a particular order of brachiopod called the spirifer, now extinct. These were creatures that lived on ancient sea bottoms and resembled clams in having two shells that opened and closed. The spirifers had winglike valves that make them strikingly resemble butterflies. But this is quite superficial. Spirifers are common as fossils. However, their relatives—other brachiopods—still live on the seafloor today.

Q *How long have butterflies been on the earth?*

A According to the fossil record, they got started during the Permian period, about 200 million years ago. They have been a very successful insect and today there are more than 100,000 species known and described with many more still awaiting discovery, particularly in the tropics. With their graceful fluttering, they have inspired much art and poetry, but other than that they are of little benefit to man. Their early caterpillar stage of development is one of voracious eating, costing us a great deal in crop loss. You can find butterflies and their cousins the moths everywhere except in Antarctica. There are superstitions about butterflies. The soul is supposed to resemble the butterfly in its flight, and to see a butterfly at night is a warning of death.

Q *I know there are frozen mammoths in Siberia. Is it true that the local inhabitants dine regularly on the meat from these beasts, which are thousands of years old?*

A They do not dine regularly on mammoth meat, but some have tried it. It is not really fit for human consumption, although dogs and other scavengers will tear away at and dismember such a carcass at those times when erosion and thawing expose these animals.

Mammoths were hunted by primitive man as short a time ago (geologically speaking) as 8000 years. We don't know the reason for their extinction, but probably it is related to climatic change as the Ice Age ended.

Q *I saw a movie about a caveman frozen in ice and thought of those large mammoths frozen in ice by the thousands in Siberia and Alaska. Why aren't more humans found frozen in ice?*

A It should be pointed out that the mammoths you mention are not found in ice, but in frozen sediment. In some cases, we know the animal fell through the ice while trying to cross a river. The body settled to the river bottom, there to be covered with sand, silt, and clay. This kind of preservation is called "refrigeration" by paleontologists. Humans preserved in this way are rare because they were lighter and also smarter, thus avoiding common traps blundered into by the ancient elephants. However, it does happen: on St. Lawrence island in the Bering Sea was found the body of an Eskimo woman who apparently was the victim of a landslide 1600 years ago. The body was in such a perfect state of preservation that scientists determined she suffered from arteriosclerosis.

Q *What is the difference between a mastodon and a mammoth?*

A Not much. There are some differences in the skull, tusks, and teeth of these two members of the elephant family. Actually, the tusks are a modification of the incisor teeth. From the fossil record, it looks like the mastodon arrived on the scene first and later developed a side branch that led to the mammoth. The mammoth in turn gave rise to the modern elephant. Both the mastodon and the mammoth survived right up to the time of man because bones have been dated at 8000–10,000 years ago. Early man hunted these beasts but was hardly responsible for their extinction; we know this because of the tremendous herds of these animals that existed. For example, the tusks (ivory) have been exported in quantity from Siberia since the Middle Ages. They probably died out because of the change in climate following the end of the Ice Age and a subsequent change in food supply. Although they are extinct, their tusks and bones will be around for a long time.

Q *I saw some kind of insect inside a chunk of yellow glass at a curio shop. The owner said it was an insect that had died millions of years ago. Is this possible?*

A There is a good chance you were looking at a piece of amber rather than glass, and if so, the insect you saw inside it was very real a long time ago, even millions of years. Amber is fossilized and hardened tree sap from coniferous trees. Insects alighting on this sticky stuff would be trapped just like on flypaper and eventually engulfed. With time, the sap would harden, preserving even the tiniest details of the insect's body. Although this type of fossil preservation is fairly rare, there is an area along the shores of the Baltic Sea that has yielded more than 2000 species of ancient insects imprisoned in amber. These include mosquitoes, flies, and gnats that are not much different from those that live today. Thus, many of the kinds of insects that bother us today and seem so lowly have existed on the earth successfully for many millions of years longer than man. Insects in amber provide another tool for the geologist in interpreting the history of the earth and the life on it.

Q *An oil well worker told me that natural gas and oil come from bugs. Was he putting me on?*

A Even if he smiled when he said it, he was telling the truth, figuratively speaking. When we think of "bugs," most of us think of flies and mosquitoes, and we doubt seriously that such creatures contributed to our oil supply. However, it is fairly well established that oil and gas have been generated from organic matter, or in other words, living creatures. Geologists theorize that the remains of ancient sea life—particularly little floating forms called phytoplankton—became incorporated in bottom sediments, and there were converted to liquid petroleum and natural gas. Other marine animals may also have contributed. Geologists argue about many things, but not about this, strange as it may seem.

Q *I saw some old statistics indicating that the United States was once the leading producer of oil. What happened?*

A We were the world's leader up to World War II, not in total reserves in the ground, but in discovering the oil we had and producing it. Our consumption of oil has continued to rise while proven reserves have declined.

What oil we have left, which is still considerable, is also harder to find because we have discovered and used up most of the easy-to-find deposits.

While the terrific deposits in the Middle East were being found and evaluated, other countries around the world were sharply increasing their demand. Japan, for instance, has hardly any oil of its own. The United States has huge deposits of coal and oil shale, which can be converted to replace natural oil deposits. This mineral wealth has not been tapped to any great extent, but we'll have to do so sooner or later. Protection of the environment during the exploitation of these resources will be vitally important.

Q *Somebody told me that when an oil well is drilled, only about half of the oil that's down there is retrieved. True?*

A Whoever told you that is being generous. Most of the time, oil companies are lucky to get a fourth of the oil in place. The oil does not accumulate in "pools" as is popularly believed, but gathers in the small cracks and interstices of the rock. The oil must be forced out of the ground. At first, the difference in pressure between the rock formation and the open boring is enough to move oil toward the well. When this difference in pressure declines, the oil will not move. However, it is possible to introduce water into a nearby well under pressure and this will force still more oil out. Even so, perhaps a third of the oil in place remains where it is. Geologists and engineers are working on ways to get all of the oil out. We need it.

Q *Why are coal and oil called "fossil fuels"?*

A A fossil is any remnant of former life found preserved in the earth's crust. Both oil and coal are indeed remains of former life. Coal is essentially the remains of plant life buried and converted. It is more difficult to imagine liquid petroleum as the remains of former life. Its organic nature is not disputed by geologists. Untold numbers of tiny organisms live in the oceans today and did so millions of years ago. When they died, they sank to the ocean bottom, there to be entombed in sediment and slowly converted to liquid petroleum. As fossil fuels, these valuable substances represent "fossil" sunlight that shone upon the earth millions of years ago and whose energy was captured and preserved by plants. The heat from burning coal is the same heat that reached the earth long ago.

Q *I was in a museum looking at crystals of minerals. Why are some crystals bigger than others?*

A Most crystals you see grew out of hot solutions under the earth. Depending on what elements are present in the solution, the atoms will arrange themselves in a particular pattern. For example, if silica (oxygen and silicon atoms) is present, six-sided crystals of quartz will form. If the solution cools quickly, small crystals result because the atoms don't have time to arrange themselves to make large crystals. Slow cooling helps to form the large crystals. Another factor is water. The more water in the hot solution, the faster bigger crystals can grow. This is because the presence of water makes atoms more mobile.

Q *I see advertisements for zircon gems. Aren't these man-made?*

A No, zircons are naturally occurring in many places around the world. The gem varieties are found in river gravels in Australia and New Zealand, but particularly in Indonesia and Sri Lanka. These are the principal source of the metallic element zirconium, which is actually more common on the earth than many more familiar metals such as copper. However, there wasn't much use for it until we got into the atomic age. Then it was discovered it was highly useful as a jacket on uranium fuel rods in nuclear reactors. The most significant source of zircon in the United States is in beach sands along the east coast of Florida. There is a good chance that some tiny crystals of zircon can be found in the soil in your backyard. Don't get too excited as it is likely to be of no economic value.

Q *Is it true that all snowflakes are six-sided but no two are alike?*

A Snow is rightly called a mineral because it is naturally occurring, is solid, and has an orderly internal arrangement of its atomic and molecular components. Being a mineral, it has a crystal structure and belongs to the hexagonal system, which does have sixfold symmetry. So indeed, many of the flakes that might fall in a single storm would be six-sided or six-pronged. But there would also be needlelike flakes and those without much symmetry at all, just like other minerals that belong to a crystal system. One oddity about snow is that flakes can fall out of a clear sky on days when there is not enough water vapor to make clouds, but enough to form on some nucleus such as a speck of dust in the air.

Q *What gives an opal its beautiful play of color?*

A Strangely enough, it is imperfections in the stone such as minute cracks and veinlets that interfere with the light as it passes through. There are numerous varieties of opal, including black opal, which is rare and expensive. Opals form from solutions of silica in volcanic areas, often in hot springs that circulate underground waters near hot magmas. Although the early Romans prized the opal and regarded it as a symbol of luck, the superstition also existed that its power brought you to strange and mysterious places. This idea may have originated because of the opal's association with the god Mercury, who served to guide lost souls through the kingdom of the dead.

Q *Where is aluminum found? You don't see it in nature like gold or copper.*

A You are right that aluminum is not very visible in nature even though it is the third most abundant element in the earth's crust, after oxygen and silicon. It is extracted from a clay (bauxite) that has survived long chemical weathering. This clay is mixed in with cryolite and heated. The process is electrolytic: the box or container (the cathode) attracts the pure aluminum while oxygen is attracted to the rods (the anodes) placed in the container and is consumed. As you know, aluminum is a light, strong metal widely used in construction and a myriad of other things, from airplanes to beer barrels. The United States and Canada are the leading producers of aluminum. The process of extraction described above is known as the Hall process, named for Charles Hall, but a Frenchman named Herould discovered an identical process at about the same time as Hall.

Forces Within

Q *I read in James Michener's book* Centennial *that the core of the earth is a solid ball of hot iron extending for 770 miles. How can this be measured?*

A Mr. Michener is only partly correct in this statement because the core of the earth consists of an outer core that is liquid, and an inner core that is solid, or behaves as a solid, and consists of both iron and nickel. We knew this from the behavior of earthquake waves or seismic waves—which travel through the entire earth—generated during a major quake. One type of wave is called the S wave; it can only travel through solids. Those S waves passing through the core never arrive on the other side of the earth. In a sense, seismic waves "X ray" the interior of the earth to show its structure, which resembles that of an onion (having concentric layers or shells). Data obtained from several seismograph stations around the earth permit the calculation of the size of this core.

Q *Is the center of the earth solid or liquid?*

A It is both. The earth has both an inner and an outer core. The inner core is solid, while the outer core is liquid. Although nobody has seen the center of the earth, we can interpret its nature by the study of earthquakes.

When there is a major earthquake, the entire earth vibrates and these vibrations pass through the center of the earth. The vibrations consist of S waves and P waves. The S waves cannot pass through liquid but the P waves can. On opposite sides of the earth, P waves arrive at seismograph stations but S waves do not. They have been stopped by liquid as they travel through the core of the earth. It is believed by scientists that the earth's core is composed of nickel and iron and is much more dense than the outer crust of the earth.

Q *The radius of the earth from the equator to the center is 3963 miles while that from Colorado to the center is 3956 miles. How can such a small difference as seven miles be measured?*

A One way to do this is by use of the gravimeter. The force of gravity, according to Newton, is known. Variations in this force of gravity can be measured at different places around the earth. The gravimeter is essentially an extremely sensitive spring that will stretch out more at the equator than it will in Colorado. Several corrections must be made, however. And by the way, this instrument is also a tool used for exploration of oil deposits because oil is rather "light." So, over an oil deposit, the gravimeter spring won't stretch so far.

Q *Is it true that the earth's crust is elastic like a rubber band?*

A Perhaps the earth's crust is not that snappy, but it does move up and down. Across all the continents we see layers of solidified sediment that could only have formed on the ocean floor and yet are now on the land, many hundreds or thousands of feet above sea level. A striking example is what has taken place during the past two million years since the Ice Age. Great ice sheets of ponderous weight engulfed the continents and shoved them down. When the ice retreated and melted, the continents rebounded. The reason we know this is true is that old lake shorelines are at different levels where they should be at the same level, indicating that some parts of the earth's crust rebounded faster than others. Today, the sediment being deposited at the delta of the Mississippi River is considerable, as the river drains half of a continent. Careful measurements show that the earth's crust in southern Louisiana is sinking down, but very slowly.

Q *Do you believe that Europe and North America are drifting away from each other at a rate of six or seven feet every 100 years?*

A It is not merely a case of belief, which presupposes some doubt, but a case of hard evidence convincing to most geologists. Take a look at a map of the world and note how neatly South America and Africa could fit together like two pieces of a jigsaw puzzle. Although strongly suggestive, the idea was rejected generally until shortly after World War II when some scientists noted that ancient rocks containing iron-bearing minerals had acted as "recorders" of the old magnetic field of the earth, much like a modern compass aligns itself in the magnetic field north–south. The directions were different on different continents, and this suggested that the continents had shifted independently of each other. Scientist Harry Hess soon came up with the idea that a new ocean floor is being created, pushing away from zones such as the mid-Atlantic ridge, and carrying the continents along like passengers on a conveyor belt, as it has been so picturesquely described. There is still much to be learned, but Hess's concept seems to be correct at the moment.

Q *Couldn't Moses's crossing of the Red Sea be explained by continental drift? The Red Sea would have been much narrower back then.*

A The "drift" just isn't that fast. There is little question today among geologists that continental drift does indeed occur, and furthermore that the Red Sea is a fine example of rifting apart from the African continent. If we assume that the Exodus took place about 3500 years ago (more if you like), the separation of the Arabian lands from Africa to form the Red Sea would not have increased during this time span by more than a few hundred feet, maybe less. The Red Sea is 100 miles wide in places. The idea that walls of water were rolled back to create a highway such as depicted in the movie *The Ten Commandments* cannot be thought of as accurate history. More likely, Moses crossed the Gulf of Suez at a place of very low water. Some have speculated that the Exodus coincided with the violent eruption of the volcano at Thera, which caused giant sea waves to slosh around the Mediterranean, temporarily withdrawing waters from the Gulf of Suez, which later returned in time to drown the army of the Pharaoh. We can't know for sure.

Q *How long did it take to form the Grand Canyon?*

A The Grand Canyon was cut by running water, the water of the Colorado River, and it took, by the estimates of geologists, about 15 million years. It is hard to believe that just water running over rocks could do this, but given enough time, it can be and is done. We are reminded of the saying of a great geologist about the little bird that comes once a year to sharpen his beak by rubbing it against a great rock; in several million years, the rock is no more. About 15 million years ago, the Colorado plateau was subject to swift elevation of the earth's crust and the old Colorado was quickened in its course and cut down with renewed energy to create this wonder of the world. As the Colorado River continues its work of downcutting, it is flowing through some of the oldest rocks on the earth.

Q *What is the worst earthquake on record?*

A It depends on what is meant by "worst." In terms of loss of life, the most disastrous quake we know of was in Shen-Shu, China, in 1556. An estimated 830,000 people died. Compare that with the much-talked-about San Francisco earthquake of 1906, which killed 500 people.

On the other hand, we can speak of the worst quake in terms of how much energy was released. Since measuring instruments first came into use, there have been several major earthquakes exceeding 8 on the Richter scale. One quake we think might have been the most intense was that in Lisbon, Portugal, in 1755. The ground shook for six minutes at one point and the city was completely destroyed. It took nearly a week to put out all the fires that started. Crewmen aboard ships in the harbor were tossed into the air. Large sea waves (tsunami) 40 feet high swept over the city. These waves traveled across the Atlantic and were still 12 feet high when they reached the island of Martinique.

Q *Do scientists agree that San Francisco could fall into the sea if an earthquake occurred?*

A The famous San Francisco earthquake of 1906 was caused by movement, in places as much as 18 feet, along the San Andreas Fault. However, most of this movement is sideways or laterally without much up-and-down movement, as with some other kinds of faults. Geologists recognize that there is now enough pent-up force in the San Andreas Fault to cause another severe earthquake should movement occur.

We doubt seriously if the city would be thrust beneath the sea during an earthquake although there might be some minor flooding in places. The chief concern is that San Francisco is a much bigger, more densely populated city now than it was in 1906.

Q *How do scientists know that a particular area was subject to earthquakes in the distant past?*

A If not too distant in the past, that is, since humans began building cities or other structures, the damage might show up in those structures. For example, the quake that struck Charleston, South Carolina, in the last century caused building damage that can still be seen today. However, even in areas without buildings, geologists would look for evidence of faulting. A fault is a crack in the earth along which rock movement has taken place accompanied by the enormous release of energy that causes earthquakes. The displacement of traceable rock units can be measured. If there is considerable displacement of tens or hundreds of feet, that would suggest several strong earthquakes having occurred, even millions of years ago. There are hundreds of thousands of faults around the world. Fortunately, not all of them are active now.

Q *I read that an earthquake in Idaho made a lake appear. How is that possible?*

A In a sense, an earthquake reshapes part of our landscape. When movement along a fault occurs, you get an earthquake. This can result in the shifting of the land surface as much as several feet at a time—up, down, or sideways. For example, after the famous San Francisco quake of 1906, it could be seen that fencelines had been offset six to ten feet or more. Suppose that instead of a fenceline, a stream course had been offset, blocking the normal flow of water. The stream water would back up and impound, creating a lake. This is what happened in Idaho. Another possibility is that the jiggling of the ground during an earthquake causes a landslide of rock and soil into a river or stream that then acts as a dam. The water becomes a lake behind it. This happened at West Yellowstone during the 1950s in an earthquake; the lake that was created is there today, named "Earthquake Lake."

Q *Geologists say they can predict earthquakes. How do they do this?*

A The state of this art is still developing, but geologists have a handle on it. There are certain clues that foreshadow an earthquake. The quake results from the buildup of pressure along a fault or fracture in deep-seated rocks. As the pressure increases, the rocks in the area start to fracture, creating openings into which groundwater migrates. The groundwater contains radon, a short-lived radioisotope. Its appearance in water wells is a signal. At the same time, sensitive tiltmeters can record upwelling of the ground. Also, seismic waves from distant sources passing through the area will slow down. Oddly enough, the farther into the future an earthquake is predicted, the more confident geologists are of the occurrence of the quake. One of the problems, once prediction becomes practical, is human behavior. For example, if we say that a quake will hit Los Angeles next September and it will be disastrous, what happens to property values? Will people leave the city in a panic? These are social questions we will have to deal with.

Q *Earthquakes and volcanoes seem to occur together. Which one causes the other?*

A An earthquake results when a sharp force of some kind is applied to the earth. The force can be artificial or natural. An atomic bomb explosion will set off a small earthquake detectable at great distances; this is how we know when testing of these weapons is occurring. We also explore for oil traps by setting off small dynamite charges in the subsurface. Natural quakes result, as you suggest, when volcanism occurs because the explosions of volcanic gas and forceful movement of molten magma below disturb the earth. However, many earthquakes, major and minor, can take place without any volcanic action whatsoever. These are called tectonic earthquakes and result when masses of solid rock slide past each other.

Q *What was the world's biggest volcanic explosion?*

A Some say the eruption of Krakatoa in the East Indies in 1883. The explosion was a big one. At one point during the two-day eruption, one cubic mile of earth and rock was shot 17 miles into the air and the explosion was heard 3000 miles away in Australia.

However, there was a bigger eruption, that of Thera in the Mediterranean Sea area about 3500 years ago. Thera was four times the size of Krakatoa. The blast darkened the entire Mediterranean, causing severe floods, earthquakes, and rains of glowing volcanic ash and cinders. In all likelihood, that blast contributed to the downfall of the Minoan civilization.

Q *How deep in the earth is the lava that comes out of volcanoes?*

A Pretty deep. Hawaiian volcanoes are among the most closely studied and monitored in the world and we know quite a bit about them. During the eruption of Kilauea in 1959–1960, the movement of the lava (or magma) upward from below could be traced by the series of small earthquakes that accompanied the movement. Geologists think the lava originated about 30 miles deep in a plastic rock zone known as the asthenosphere. The molten lava eventually reached the surface, with spectacular flows and lava fountains. This lava is mainly a dark lava called basalt. Some of it may originate even deeper.

Q *I was told that there are hot springs in Iceland. How can a place so cold have hot springs?*

A Hot springs can be found around the world in such places as New Zealand and the western United States, climates different from Iceland. What is interesting is that the hot springs and what lies below them gave birth to a land as cold as Iceland. These hot waters come from molten rock sources far below the surface that provide the raw material that is ejected from volcanoes. In the remote past, eruption of this molten rock material on the ocean floor caused a buildup of basalt, eventually reaching the surface and forming the island we call Iceland. The molten rock and gases are still stewing beneath Iceland as the hot springs attest. The emergence of the volcano Surtsey during the 1960s south of Iceland shows how new land is created. Our 50th state, Hawaii, was created in similar fashion, as were most of the islands in the Pacific.

Q *Is there a chance that Mt. Saint Helens will erupt again any time soon?*

A There is always that possibility, although usually periods of eruption are followed by long periods of dormancy. When Mt. Saint Helens erupted on May 18, 1980, it was the first time it had done so in 123 years. Keep in mind that Helens is only one of a string of volcanic peaks along the Cascade Range that have the potential to erupt. Among these are Mt. Shasta, Mt. Rainier, Lassen Peak, and also Mt. Hood and Mt. Baker. Lassen Peak last erupted in 1915. Geologists have found that these are actually newer volcanoes built upon much older volcanoes dating back more than 50 million years. The "new" volcanoes like Mt. Saint Helens are only about one million years old.

Q *There is a new book that predicts the earth will come to an end in 1999. What do you think?*

A We have read the book you mention. It purports to offer scientific evidence from geology, but we couldn't find it. The author states that there has been a continuous increase in the number of earthquakes as a forerunner of the disaster. More likely, there is an increase in science's ability to detect and identify earthquakes. Thus, there would only *appear* to be more earthquakes now.

This book also states that legends and ancient writings forecast an end to our civilization in 1999. Ancient writings are voluminous and varied. If you select the writings to prove your theory (whatever it is), you can "prove" almost anything. The earth has existed for 4½ billion years and there is no scientific evidence that anything special will happen in 1999. Of course, we can destroy ourselves by atomic war or pollution of our environment at some time in the future, but science obviously has no way to pinpoint the year of such an occurrence, if it were to take place. We might mention that in 999, there was considerable hysteria in places around the world because people thought the end would take place. Nothing happened. Probably the same thing will happen as we approach 2000.

Surface Processes

Q *I've seen both the Rocky Mountains out West and the Appalachians in the East. Why are the Rockies so much higher and more rugged?*

A Both mountain ranges are millions of years old. But the Rockies are much younger and so the forces of weathering and erosion have had less time to wear them down following the uplift of the earth's crust that created them. Actually, the Appalachians of today are a sort of reincarnation of earlier mountains. About 200 million years ago, the Appalachians rose up and may have been as impressive as the Rockies. With time, they were eroded down to a plain. This took millions of years. Then, a further upwarping set rivers and streams to work dissecting the "roots" of the old Appalachians to bring harder rocks into relief as higher places while wearing away the surrounding softer rocks. The present topography is the result. The earth is so old there was time enough to make and destroy twenty such mountain chains, but that hasn't actually happened.

Q *I know almost anything is possible these days, but is it true, as I've heard, that some people still believe the earth is flat?*

A As incredible as it seems, IFERS, The International Flat Earth Research Society, with headquarters in California, is an organization whose members hold the belief that the earth is flat. Furthermore, their doctrine considers that continental drift is actually the land and oceans being shaken apart by God. They maintain that the earth is not a planet and does not spin like a ball, being instead an infinite world, without end. The Society was founded in 1800. Flat earthers define themselves as seekers of the truth regarding the geophysical earth. They consider theory as imaginary and focus instead on knowledge that is "provable." Science does not support their claims.

Q *Are other whirlpools, similar to the one below Niagara Falls, found around the world?*

A There are several places in the world's oceans and shorelines where whirlpools may be found. In the open ocean, they are simply large-scale eddies. The Sargasso Sea is one. In coastal areas such as those of Scotland, Norway, and Japan, a whirlpool can be generated owing to the configuration of the coast, bottom topography, and the action of incoming and ebbing tides. Some smaller vortices may on occasion suck down floating objects, but the idea of a whirlpool sucking down large vessels is pure fiction. In river bends, an upward swirling current of water (called a kolk) may occur, but this is unlike the whirlpool in the Niagara River.

Q *What is the largest lake in the world?*

A A tough question because there are freshwater as well as saltwater lakes. And are we talking about largest in terms of surface area or amount of water contained in the lake? Without question, the inland saltwater lake of the Caspian Sea, with its 4000 miles of shoreline between Europe and Asia, wins the prize as the biggest lake. If we are talking about freshwater lakes, Lake Superior is the largest lake in terms of surface area, but it does not contain nearly the amount of water as does Lake Baikal in Siberia. Most of the present lakes of the world are the result of meltwaters remaining when the ice retreated after the Ice Age.

Q *Giant tidal waves called tsunami are said to reach heights of 200 feet. This seems unreal.*

A It is unreal. On rare occasions they may reach 120 to 130 feet, but an "ordinary" tsunami of, say, 50 feet is impressive enough. Unlike regular waves, which are produced by the wind, tsunami are produced by a blow to the ocean floor such as an earthquake or volcanic explosion. Following such an event, the tsunami travels with great speed across open ocean (at several hundred miles per hour) and builds up to great height as it enters shallow coastal areas. Oddly enough, far out at sea the tsunami waves are hardly noticeable, being only two or three feet high.

The Pacific Ocean is more prone to produce tsunami because of the earthquake activity there. Perhaps one or two a year occur. In 1883, the eruption of Krakatoa generated tsunami that struck the shores of Java and Sumatra, sweeping more than 20,000 people to their deaths. In 1946, tsunami hit the Hawaiian Islands, killing 159 people and causing $25 million in damages. It should be noted that tsunami are not tidal waves as they have nothing to do with tides.

Q *Why is the Dead Sea so named?*

A Although it has other names, for example, the Salt Sea or the Sea of Zoar, the Dead Sea is probably the most descriptive name for the lake into which the Jordan River flows. Its waters have a salt content of 25%, so that fish are not able to live there. Being the lowest body of water on the earth's surface, the Dead Sea has no outflow and instead evaporates as new waters are added from the Jordan and other rivers as well as from underground springs.

Dead Sea minerals have supported an industrial production of potash, salt, and so on. Tourists come for the mild winter climate and historic associations such as Masada and the caves where the Dead Sea Scrolls were discovered. There are no modern cities, however, anywhere around the shoreline.

Q *Is there scientific proof that a worldwide flood occurred in Noah's time, and an ark housed the only life on the earth?*

A More than one billion cubic miles of water would have to be added to the world's oceans to cover all of the highest mountains. Science knows of no source for such vast amounts of water, nor any place to put it after the flood receded. There is a space problem in the ark. You would need to house more than 43,000 animals, to be cared for by only eight people for nearly a year, plus food and water. The ark wasn't that spacious.

If the story were in fact true, then all humans are descended from Noah's family, including such diverse peoples as African Pygmies, blond Swedes, and Alaskan Eskimos.

Big, but local floods did occur in Noah's area; science has evidence of these. This may have been the basis for the story of the flood and still earlier large floods written about in ancient texts.

Q *I recently visited Carlsbad Caverns. Just what causes caves to form?*

A Most caves are formed by the dissolution of limestone rock. Water trickles down into the subsurface and becomes charged with organic acids when it comes in contact with plant material. It then reacts with the calcium carbonate of the limestone and a cavity develops. Over a period of thousands or even millions of years, a cave is formed. Water dripping from the ceiling of a cave creates the well-known stalactites or "icicles" as the water evaporates and deposits calcium carbonate. Caves such as Carlsbad Caverns and Mammoth Cave have not been fully explored. Caves were important as places where early man lived.

Q *Can water in the ground be located by water witching?*

A This is what is called dowsing: a forked stick is held by a person who walks over an area in which water is hoped to be found. When the stick seems to be pulled downward, water in the subsurface is indicated. In many cases, drilling reveals that water is present. This is not surprising because water should be found anywhere you drill if you go deep enough, even in the middle of the Sahara desert. That a stick will locate water is nothing more than a superstition, but believed by many. We would suggest a more logical course if you wish to have a well. Consult state or federal agencies that have groundwater maps of your area; they will be happy to advise you as to your prospects, including how deep you may have to drill.

Q *Is is true that a landslide roars down the slope at great speed without touching the ground?*

A Oddly enough, it is true and you can verify it for yourself. If you drop a book on a table, notice that air escapes out from under the book and disturbs light objects nearby. In the same way, a mass of millions of tons of rocks and soil moving down a slope will cause air to rush out from under it at velocities of 50 or 60 mph. There will be a cushioning effect of the air before it can escape. Thus, the landslide may be riding on a layer of compressed air a few inches above the surface of the ground. This also accounts for the speed at which they can move—up to 100 mph. The proof lies in careful examination of the track of the landslide. There will be blades of grass and small pebbles that remain undisturbed.

Q *How can a solid rock turn into clay?*

A It is a perfectly natural process called weathering. It goes on all the time. However, it is a remarkably slow process taking thousands, even millions of years.

You might think of the surface of the earth as a gigantic chemical laboratory where the minerals of rocks, which are chemical compounds, are reacting with water, oxygen, and carbon dioxide in the atmosphere. The result is rock decay. The process is little different from the rusting of iron objects left outside, for example. Incidentally, the moon rocks brought back by astronauts contain very fresh-looking minerals, even though they are of great antiquity. This is because the moon has no atmosphere.

Q *What do the tiny sand grains that make up the beach look like under a microscope?*

A It is a magical world at 200× magnification. Each little grain, smaller than the head of a pin, becomes a very large block of material. They are transparent in many cases and show a variety of shapes. The most common grain is that of quartz, revealing impact scars where it has collided with other grains. Most interesting are the "accessory" minerals including such exotic ones as tourmaline, zircon, garnet, and hypersthene, which display a variety of colors, shapes, and crystal forms. The numbers and order of abundance of such minerals occurring in sands and soils have been instrumental in solving many murder cases by associating the suspect with a particular site. In this way, science can help bring criminals to justice.

Q *Where does sand on a beach come from?*

A Streams inland from a shore are busy weathering and eroding rock material. This material slumps into a stream or river and is transported downslope, often for many miles. During this transport, size reduction of particles takes place due to abrasion and impact. A mixture of particles arrives at a beach on the ocean or a large lake. There, the waves' to-and-fro action sorts out the finer particles of silt and clay from the sand-sized particles, which are left on the beach and accumulate to form the shoreline. Finer particles are swept out into deeper waters. Under a microscope, these bits of sand are seen to be of different mineral origin, although most of it is quartz. In tropical areas, the sand is made up of fragments of coral. Sand is a size term and does not indicate composition. Sand consists of particles between 2 mm and 1⁄16 mm in diameter.

Q *I was on vacation in Florida and saw heavy mining equipment scooping up tons of sand. Sand is cheap, so why this expensive operation?*

A It is probable you were witnessing a heavy-mineral extraction operation. Commonly, the sand along a beach is mostly the mineral quartz, which has little value. But also present are grains of other minerals that, if there are enough of them, can be a valuable resource. For example, rutile and ilmenite, in quantity, constitute an ore of the metal titanium. Because of their higher specific gravity, nature concentrates these minerals along beaches and streams in what are called placer deposits. In the extraction process, these minerals are further concentrated by gravity or electromagnetic means. There is a great variety of heavy minerals, including magnetite, kyanite, tourmaline, garnet, and zircon.

Q *So many movies show the villain getting sucked down by quicksand. What is so special about this kind of sand and where is it most often found?*

A First of all, there is no special sand involved. Any sand can become "quick." When you are walking along a sandy beach, the sand supports your weight because all of the sand grains are in mutual contact and have a bearing strength. However, if there is present a current of water (such as from a spring) that is passing upward through the sand, then the grains lose contact and the sand–water mixture behaves like a liquid.

Quicksand does not suck you down. The sand–water suspension actually has more buoyancy than water alone because it is denser. The reason why animals, and some people, have drowned in quicksand is the panic and struggling that ensue. Quicksand is more often found in swampy regions because there are areas of possible intersection with the groundwater table, and a greater likelihood of rising waters from below. If caught in quicksand, remain calm and yell for help.

Q *I was studying a map of the world and noticed that the deserts seem to form a band north and south of the equator at about 20° latitude. Is this just a coincidence?*

A No. This pattern results from winds moving from higher latitudes toward the equator. As they do so, they pick up moisture but are unable to release it as rain because as they get closer to the equator they increase in temperature and thus can absorb greater amounts of moisture. At the same time, these winds are rising higher and higher, as hot air is well known to do. In this way, the air passes through the regions of 20° latitude without dropping moisture, and thus these areas become deserts. As the winds arrive over equatorial regions, they come in contact with colder regions of the atmosphere, cool down, and release all the water to help create the tropical rain forests near the equator.

Deserts also form in areas where a mountain barrier obstructs rain-bearing winds, and also areas where, as in the center of large continental masses, the source of water lies at a great distance. There are more desert areas in the world than most people realize.

Q *Can the wind in a hurricane be strong enough to cause thin sticks to penetrate a plate of steel?*

A It is possible, but don't imagine the steel plate to be several inches thick. The power of the wind can only be appreciated when a person experiences it. It is awesome. Even "moderate" winds of 70–80 mph can set people rolling in the streets. Ripley reports a straw found impaling a glass plate after a hurricane.

In our own experience, we have observed thumbnail-sized pebbles embedded in wooden fence posts in Texas.

Some were about three feet above the ground. Mute testimony to the power of the wind.

Q *Why does the leaning tower of Pisa lean?*

A The structure started to lean during construction. Although the foundation was dug down to ten feet, they did not reach bedrock for a firm footing and it began to settle soon after. The tower was started in 1174 by Bonanno Pisano and not finished until almost 200 years later. Its present height is about 185 feet, and the tower leans almost 17 feet from the perpendicular. What is amusing is that instead of starting all over again, during construction they added stories at an angle to straighten the tower. Instead of the leaning tower, it should be called the crooked tower. During the 1960s, cement was introduced into the foundation to strengthen the tower but it keeps leaning more and more and is still threatened with collapse. Perhaps it should be called the tired tower of Pisa.

Q *I have visited the Petrified Forest in Arizona. How could a whole forest turn to rock?*

A One of the myths of the Petrified Forest is that the trees that are there today actually grew where they are now seen, and even advertisements for the Petrified Forest show trees in a standing, growing position. This is false. Geological research shows that the trees actually grew in Colorado about 180 million years ago. They died, fell, and were transported by rivers into the northern Arizona area, there to be buried in sand and silt. With time, circulating groundwater loaded with silica replaced the woody fibers with a molecule of silica on a volume-for-volume basis, thus preserving the woody character of the trees, even the bark. The most common form of silica is quartz, which makes up a large portion of the earth's rocks. Smaller fragments of petrified wood can be found all over Texas, so it is not a rare phenomenon, although the preservation of entire tree trunks as at the Petrified Forest is unusual.

Q *Are the catacombs of Rome that the Christians hid in natural formations or were they dug out?*

A The underground passageways and small galleries under Rome were artificially constructed by excavation of a rather soft layer of volcanic rock. If your question implies that they were dug as hiding places to escape persecution, that is not true. During the first three centuries, both Jews and Christians constructed them as burial places, or in effect, large underground cemeteries. In places, there are several descending levels, and the catacombs became very elaborate and extensive. Thus, they also did indeed serve during times of persecution as places of refuge from Roman soldiers. After the third century, they were mostly abandoned and entrances collapsed over many years, so that today, many passageways and burials remain undiscovered. Catacombs are not restricted to the area of Rome, but are found in many countries around the Mediterranean area.

Q *I saw a large mass of rock with parallel sets of scratches in it that someone said was caused by ice. Isn't this another case of a weak explanation about something we know nothing about?*

A To the contrary, these scratches, or striations, are well understood by geologists. It is only baffling if you think of ice as cubes in a freezer or a thin layer to slip on while walking on a sidewalk. We see these scratches on bedrock in areas we know were once glaciated. Imagine ice covering this area on a large regional scale and with thicknesses of several thousand feet. We know this to be true because of the great drop in worldwide sea level, the oceans being the only logical source for all the water to make the ice. These ice masses move slowly and plastically and pick up stones, which are gripped by the ice at the base of the ice sheet. As the ice moves along, the stones held by the ice do the scratching. Geologists studying modern glaciers have observed such striations forming. Those formed long ago probably developed the same way.

Q *Is it true that ice sheets more than one mile thick used to exist in the state of New York?*

A It is true, and in places the ice was more than a mile thick. Also, the ice sheets extended far beyond New York both east and west across the United States. They began to melt about 12,000 years ago. We know the extent of these great ice sheets because during their advance, they picked up enormous quantities of sand, silt, clay, and even large boulders. When the ice melted, they left this sediment behind on the ground. By mapping this glacial debris, we know how far it reached. We have a good idea of the thickness because the ice caused the earth's crust to sag down due to the weight. Great ice sheets still exist in Greenland and at the South Pole, and they are certainly more than a mile thick.

Q *I am confused. Some say a new ice age is coming while others declare the earth is warming.*

A We have seen the same contradictory reports. There is insufficient information to favor either theory on a long-range basis. We simply do not know. Good data on climatic changes go back only a couple of hundred years, which is not long enough to assess long-range changes on the comings and goings of ice sheets, which progress over many thousands of years. There have been rather severe winters lately, it is true, but this may just be due to short-term fluctuations. On the other hand, there seems to be some melting of the Antarctic ice cap and retreat of the big Columbia glacier, but this does not necessarily signal the melting of all the ice on the earth. We shouldn't get too "hyper" about sensationalistic talk. Whichever outcome occurs—hotter or colder—it is going to take a long, long time.

Q *What would happen to the earth if all the ice at the poles and Greenland melted?*

A Sea level would rise around the world, by perhaps as much as 200 feet. If it happened all at once, it would be the worst catastrophe of all time because the world's shorelines would be inundated. Most of the population of the world lives along shorelines. The geography would be changed. For example, the entire Florida peninsula would be under water. Fortunately, this won't happen. Still, the ice is melting slowly and world sea level is rising, but measured in inches per century. During the past two million years there have been significant changes in sea level. Many places that once were human habitation sites are now under water. Conversely, as ice sheets grow, sea level falls. This would create new land areas and permit migration of animals and man across land bridges. This is how man came to North America from Asia across the Bering strait.

Q *Why is our government spending so much money to drill holes in the Greenland ice cap?*

A Other countries besides ours are chipping in the money, especially Denmark. In drilling through this very thick ice, cores of the ice are obtained for study. The deeper ice is many thousands of years old and can actually be dated because small air bubbles, trapped in the ice, represent the composition of the primitive atmosphere of the earth. Since this air contains carbon dioxide, it can be dated by the carbon-14 method. It tells us how the atmosphere has changed over many thousands of years (it has not changed much despite man's polluting of the air). Also, layers of dust particles represent major volcanic eruptions in the earth's past and this can be related to climatic change. The more we understand the earth and its processes, the better we are able to control the forces of nature for the benefit of mankind.

Miscellaneous Phenomena

Q *Why is mathematics called the "queen of the sciences"?*

A Well, all scientists have a need at some time or other to measure things in their work and this requires at least a fundamental understanding of mathematics. Another explanation is that a pure mathematician need not know any other science to carry out his work, whereas physicists, chemists, geologists, and all others must know mathematics before they can work in their fields. It was the Babylonians who first established a science of mathematics, but the Greeks made strong contributions, especially Euclid, who wrote a book called *Elements* in 300 B.C., laying down the foundations of math and geometry. It is estimated that Euclid's book, in human history, is second only to the Bible in its circulation. While math is a tough struggle for some students, it is one of the most useful areas of learning.

Q *I can't see much use in the centigrade scale. Why push for this when all it does is create confusion?*

A It is part of the effort to convert over, and bring into popular use, the metric system. The centigrade scale is certainly logical, with 0° being the freezing point and 100° the boiling point of water. In contrast, the Fahrenheit scale has awkward numbers (32° for freezing and 212° for boiling water). We agree there is confusion. If we had grown up using the centigrade scale, we would complain about the Fahrenheit scale if it was being foisted upon us. The metric system has advantages because it is based simply on 10, and most of the rest of the world uses it. Note that we use it in our money—ten cents equals a dime and ten dimes equals a dollar. In contrast, the English system of money is a real mess, with their farthings, quids, and pounds, which follow no logical pattern.

Q *With all the computers we have today, I sometimes wonder who was the person who invented the computer.*

A There is no single person who is responsible for the computer. Perhaps the forerunner was the simple counting device known as the abacus, which has been around for more than 2000 years. Even before that, due to mathematical necessity, various crude computers or methods of computation were in use. For example, an army about to go into battle would be concerned about casualties. All soldiers would place one rock in a pile. After the battle, each surviving soldier would remove one rock. The remainder represented the losses. If you are thinking about electronic computers, the first was made in 1946 and called ENIAC. It was designed by J. P. Eckert and J. W. Mauchly of Philadelphia. It was a bit clumsy, containing 18,000 vacuum tubes. Later, UNIVAC was built and 20 similar types followed within a few years. The whole field then expanded with better and better computers, as we know, until the present dynamic industry was reached. We are by no means at the apex of computer growth and development.

Q *The speed of light is about 186,000 miles per second. How were scientists able to measure the velocity of something so fast?*

A In 1676, an astronomer named Roemer was observing eclipses of Jupiter's satellite Io, when Jupiter was at its nearest point to the earth. He cleverly deduced that 6½ months later, when another eclipse of that satellite would take place at Jupiter's farthest point from the earth, he could get an idea of how long it would take light to cross the diameter of the earth's orbit around the sun. The delay in arrival time from that theoretically predicted was 22 minutes, which translated to about 180,000 miles per second as the speed of light traveling to reach the earth. Corrections were later made and other methods were also used here on the earth to determine the speed of light. In 1850, the Frenchman Fizeau and others used two mirrors separated at some distance, one of which was rapidly rotating and the other stationary. Light from the rotating mirror traveled to the stationary mirror and back to the rotating mirror. The amount of rotation of the mirror during the beam of light's journey permitted determination of the speed of light. Despite this incredible velocity, space is so vast it requires years for light to travel from one star to another.

Q *Does lightning never strike twice in the same place?*

A This is popular folklore but it is not true. Indeed, if a place is struck by lightning once, there is a very good chance it will be struck again. High buildings such as the Empire State Building in New York City have been struck by lightning many times. Once electricity has accumulated in the atmosphere, you need only positive and negative charge centers to cause the discharge of electricity in the form of lightning, which is really just a gigantic electric spark. This is not to say electric storms are not awesome or dangerous. Ancient peoples regarded thunder and lightning as examples of their god's anger or handiwork. In some African tribes, fires caused by lightning would be left to burn and people struck by a bolt would not be helped for fear of interfering in the work of the god. Even today, superstitions persist. It is not uncommon for mothers to tell children that thunder is God moving furniture in heaven.

Q *How does a battery-powered flashlight work?*

A The battery converts chemical energy into electrical energy by stripping atoms of their electrons and causing them to flow, which is of course an electric current. The stream of electrons passes through the filament in the bulb and heats it, causing it to glow. The concave reflector in the flashlight directs the glow so you can see better. We know that three billion, billion electrons flow through an ordinary 60-watt bulb every second. The metal tungsten makes an excellent filament for the electrons to pass through because it has the highest point of any metal. For example, at a certain temperature, iron will boil while tungsten will still be a solid. Thus, tungsten can resist the high temperatures generated.

Q *Has anyone succeeded in inventing a perpetual motion machine?*

A Not that we know of. Over the centuries, many have tried to build such a machine, i.e., one that would keep working indefinitely, if not forever, without drawing upon any outside energy source. It has held as much fascination as did alchemy in the Middle Ages, which sought a way to change lead into gold. The most common type of perpetual motion machine was an overbalanced wheel; others relied on a continuous circulation of water. All such efforts fail because they violate the scientific law involving the conservation of energy. Yet there were many hoaxes and frauds. In 1870, John Keely raised a lot of money, claiming a motor he had invented could power big ships across the Atlantic. Some devices such as a self-winding watch may appear to be a form of perpetual motion, but there is an outside energy source in the movements of the wearer.

Q *Why is ice so slippery?*

A Ice has several unusual properties, one of them being that it melts when subjected to pressure. Your foot on ice is such pressure, and a film of melted ice—water—reduces the amount of friction and thus sliding can occur. This is also the reason why an object placed on ice can become embedded in the ice and melt its way through the ice by melting ahead of and refreezing behind the object.

About 99% of all the ice on the earth is found in the Antarctic and Greenland. It is most fortunate that this ice does not melt; if it did, sea level would rise 200 feet and change the entire configuration of the continents.

Q *What causes the loud noise or boom when a plane exceeds the speed of sound?*

A As long as a plane is flying at subsonic speeds, the disturbance to the air is well in front of the aircraft. As the plane builds up to Mach 1 (the speed of sound), a sharp pressure rise occurs in front of and tangential to the plane and creates a shock wave that manifests itself as claplike thunder. In a sense, the air molecules crowd together and collectively impact. It is interesting that the pilot of the plane does not hear the sonic boom, although on the ground, it not only is heard, but can break windows. The effects of supersonic speeds were first described by the Austrian physicist Ernst Mach in 1881.

Sound travels at a speed of about 1100 feet per second at sea level, and this is why you might see someone say, chopping wood at some distance, and see the ax strike before hearing the sound.

Q *A department store wants me to take a lie detector test for employment. How accurate is the lie detector?*

A The first thing we should understand is that the so-called lie detector (or polygraph) does not detect lies. It records human emotional responses in terms of changes in heart rate, perspiration, and so on. These responses must be interpreted by an operator skilled in the use of the polygraph, and not everyone is. Let us suppose a person sincerely believes he has seen a flying saucer and is given a lie detector test. Although the flying saucer does not exist, the lie detector will indicate the person is telling the truth. Sensationalist writers seize upon this as "proof" that flying saucers exist. Not so. Another thing to be considered is that we all have guilt feelings about events in the past, and a question may touch upon these feelings and create an indication of guilt that can confound even an expert. While the lie detector may have its uses, it is by no means a magical instrument.

Q *Why does water boil?*

A When you apply heat to a container of water, the heat energizes and excites the molecules of hydrogen and oxygen that make up the water. These molecules start crashing into each other. There are millions of these collisions, which release energy in the form of heat, and that is why the water gets hot, in the same way that if you beat a piece of metal with a hammer, the metal gets warm.

As the agitated molecules bounce around, they tend to separate from one another and occupy a larger volume. When the density decreases enough, bubbles will rise through surrounding water that is more dense. The result, of course, is steam. If the steam, as water vapor, continues to rise high into the sky, it will cool, condense, and regain its liquid form. If you are outside when this is happening, be sure to have an umbrella.

Q *Why doesn't water have any calories?*

A There is nothing in water that provides a source of energy for the body in the sense that something like carbohydrates can be "burned" by the body to create energy. Yet as we all know, without water you will not survive long. Water in our bodies permits, and participates in, the chemical reactions essential to life. Throughout history, people have recognized the magical qualities of water, or for that matter, any liquid. Water is the universal detergent. Such ceremonies as baptism represent the cleansing of the soul in much the same way as water cleanses the body of dirt. Water is associated with life and the renewal of life. Thus, the Spanish explorer Ponce de Leon sought the fountain of youth in

Florida and landed in St. Augustine. It is ironic that people in the twilight of their lives go to St. Augustine to await death in the same place he sought eternal youth.

Q *How do they get mercury to stick to glass to make mirrors? It doesn't seem to stick to anything.*

A In the old days, they would lay out a flat sheet of tin foil and spread mercury on it. The tin and the mercury would form an amalgam or alloy that allowed it to stick. Then very carefully a sheet of glass would be lowered until it made contact with the mercury. Weights were then placed on the glass to squeeze out the excess mercury. Prior to this, the only mirrors were made of polished metal. In the Middle Ages, thin sheets of reflective metal were used as backing on glass. In the 19th century, Liebig, a German chemist, discovered how to make mirrors by silvering. This was better than using mercury. An ammoniacal solution of silver is used in conjunction with Rochelle salt or some other compound.

Q *How do oysters form pearls and what is the largest pearl known?*

A Actually, oysters do not form the kind of pearls you are thinking of. Oyster pearls are purplish and without luster and therefore of little value. True pearls come from other kinds of mollusks (those that have two shells or valves that open and close) found in many parts of the world including the Pacific and the Caribbean. If a foreign particle such as a grain of sand gets inside the mollusk, it acts as an irritant and the animal secretes calcium carbonate around it in concentric layers; this becomes the pearl. A thriving business today is in cultured pearls where some foreign substance (often tiny seed pearls) is deliberately introduced inside the animal and the completed pearl harvested a few years later. The largest gem-quality pearl in terms of weight is one of 34 carets owned by the French Crown. Another pearl, but not of gem quality, weighs 620 carats. Incidentally, there are freshwater mussels that produce high-quality pearls found in such rivers as the Mississippi.

Q *The ancient alchemists tried to turn lead into gold. Did they ever succeed in anything important?*

A Yes. They laid the foundation of the science of chemistry. In medieval times there were many who called themselves alchemists but might be better called mystics or charlatans. The practical alchemists were for a long time obsessed with changing base metals such as lead into gold. They spoke of a philosophers' stone that would achieve this. Actually, the philosophers' stone might not have been a stone at all, but rather a substance or even a laboratory procedure. This serves to illustrate that alchemists wrote their experiments and books in cryptic, symbolic language, perhaps to guard "trade secrets" or in fear of being charged with heresy, a serious crime in those days. Alchemists designed much of the laboratory equipment used today, such as beakers, flasks, retorts, furnaces, and distillation apparatus. They recognized and described several new classes of chemicals such as caustic alkalies. Gradually, alchemy merged into chemistry and became systematic. It is interesting that transmutation of one element to another, dreamed of by early alchemists, has been accomplished by modern nuclear physics.

Q *How is kerosene produced?*

A It is one of the components of crude oil from wells. The oil, as it comes from the ground, is heated. As this is done, the most volatile hydrocarbons (e.g., gasoline) are driven off. After that comes kerosene and then the thicker oils and greases. When drilling for oil became routine more than a century ago, it was to obtain the kerosene for illumination in lamps. This replaced whale oil and we assume the whales rather liked the idea. It is amusing that the gasoline, which had to be drawn off first, was troublesome and dangerous. The early drillers dumped it in creeks where it caught fire frequently. At the turn of the century, with the development of the internal combustion engine, a use was finally found for gasoline. Even so, many people said gasoline use in automobiles would cause them to blow up.

Q *Why does some soap float, and how is soap made?*

A The soap that floats has small air bubbles in it, making it lighter than the water. The bubbles are retained in the soap during the manufacturing process by aeration, which involves rapid chilling of the soap material. Although soap chemistry is very complex, it involves basically a reaction between some kind of fat and a caustic solution (such as sodium hydroxide). This produces soap and glycerol. Part of the soap manufacturing process is to separate the glycerol.

Early Roman writers refer to the first soaps being made by boiling together goat tallow and wood ashes, so soap has been around a long time, but not in mass quantities. Even as late as colonial times in America, each home made its own soap. With a general lack of soap until the 20th century, people did not bathe much, and one account we read tells of a young woman taking a bath for the first time and finding out that it wasn't too bad once you were immersed. Kids in those days must have been happy: no Saturday night bath.

A final point—shaving cream is often mostly cold cream and not true soap, and the chief beard-softening ingredient is . . . water.

Q *Who invented ice cream, and when?*

A Nobody knows who. It is known that at the time of the Roman Empire, they added flavor to ice and licked away at that. Marco Polo is said to have returned from China with a recipe for making ice cream using milk. Once it was discovered that adding salt to ice lowered the temperature below that of ice, and mechanical refrigeration came into being, the ice cream industry prospered. Chefs in Europe kept their ice cream recipes a secret so that only the nobility would have it. This scheme did not work, which served them right, and soon others had ice cream. Most ice cream consists of milk, cream, nonfat milk solids, sugar, and sometimes eggs. Besides calcium, ice cream contains proteins and vitamins A and B. The most popular flavor is vanilla, even though more than 100 flavors are known.

Q *Is is true that the Chinese discovered gunpowder?*

A Maybe not. The English, Arabs, Hindus, and Greeks say they did. We know that very early in history the Greeks concocted a sticky mixture of sulfur, pitch, and oil that was called "Greek fire" and terrorized their enemies when thrown at them because it not only stuck, but its flames were hard to put out. Perhaps this was the beginning of gunpowder, or black powder, as it was otherwise known. Gunpowder is mainly saltpeter, or potassium nitrate, which, when blended with about equal amounts of charcoal and sulfur, burns fiercely. If confined in a container of sorts, there is an explosion. Perhaps 600 to 700 years ago, gunpowder thus became something to use to blow things up. Some historians think the advent of gunpowder signaled the end of the large castles of the Middle Ages because the walls could be blown up. Without a doubt, gunpowder changed not only warfare, but also human history.

Q *How did the ancient Egyptians make papyrus?*

A Actually, they made paper out of the papyrus plant. It grew in the delta of the Nile as a reed with a long slender stalk. The Egyptians sliced the stalk, or stem, lengthwise to form strips and placed them side-by-side. They then placed other such strips at right angles to form a sort of weave. After soaking in water, it was beaten with a mallet, dried out, and smoothed. And you had a sheet of paper. Several sheets were joined together to form a roll. The person or persons responsible for this process are not known. The Egyptians used the same papyrus reed to make boats, sails, and even sandals. It was also edible either raw or cooked, and the Egyptians ate a lot of it. It was a very handy plant—if you got hungry, you could eat your own shoes.

Q *The use of cosmetics is such a big thing nowadays, what did women do before there was a cosmetics industry?*

A Woman, and often men, have been using cosmetics since civilization began. Probably the roots of the cosmetics industry are in ancient Egypt. There are murals 4500 years old showing women with hairdressers, and using combs and polished metal mirrors. The tombs of that era and later contained urns and jars of aromatic oils and unguents. When some of these were unearthed after 4000 years of burial, the scent was still detectable. The Egyptians imported many of the ingredients from Arabia. It is known that women of many centuries ago used perfume, eyeliner, and skin softeners. Some things don't change. However, one thing they didn't have was the modern permanent wave. This was introduced in about 1906 and was popular, even though a woman had to sit for 10 to 12 hours, endure a fair amount of discomfort and pain, and pay a lot of money for it. The growth of the beauty shop industry came at about the same time because the home did not have the equipment needed for the permanent wave.

Q *Why does science dispute the Bible?*

A Science does not dispute the Bible except where it is claimed that the Bible represents scientific truth. The Bible is a source of religious faith and does not require the exercise of reason for its believers to follow its teachings. In that context, science is not involved and does not care to be involved. It is only when enthusiasts of the Bible claim there is scientific evidence to support biblical assertions that science enters the picture. Each claim should be judged on its own merits. The statement that the worldwide flood of Noah as depicted in the Old Testament is supported by geological evidence is not true. Quite the contrary. Yet other aspects of the Bible may have a scientific foundation, subject to the data and testing of science. It should be noted that many scientists accept the Bible as part of their religious life—a source of inspiration, but not a scientific textbook.

Q *Is there any possibility that science might some day build a time machine to travel into the past or future, or is it all fantasy?*

A As intriguing as the idea is, we would have to say it is fantasy. The past involves events that had a beginning and an ending. It is a one-way street. Otherwise, we would have to allow that, say, Julius Caesar, somewhere in the universe, must perpetually be having his diapers changed, cross the Rubicon again and again, and he forever stabbed by Brutus. It does not seem reasonable, or even fair to Caesar. To travel into the future is to view something that does not even exist.

However, one aspect to this fascinating topic is that we can view the past right now. We look out at stars many light-years distant. We see them not as they are now, but as they were when the light left them. For example, a star eight light-years distant is seen as it was eight years ago. An astronomer on a planet 2000 light-years away, had he a magical telescope of superpower, might see Caesar cross the Rubicon. A mythical astronomer five billion light-years away might see the earth itself come into existence.

Q *When I see the Goodyear Blimp at football games, I wonder when the first blimp was invented.*

A For more than 200 years, perhaps longer, man had tried to fly in balloons but was largely unsuccessful because there was no suitable engine to propel the airship. In 1851–1852, the Frenchman Henri Giffard built a ship equipped with a steam engine. It worked. He flew his 144-foot-long vessel over Paris at the breathtaking speed of 6 mph. Since that time, many dirigibles have been built in many countries. Most of the early airships were filled with hydrogen, which has great lifting power but is highly flammable, as witness the *Hindenburg* disaster in 1937, when that zeppelin burned at Lakehurst, New Jersey, and 36 people died. Helium is the safer gas now used. By the way, the *Hindenburg* could cruise at 78 mph, a big improvement over Giffard's 6 mph. Goodyear has been a major builder of airships in the United States since 1911.

3
LIFE ON THE EARTH

Evolution

Q *Can you clarify the debate going on between evolution and creationism?*

A In science classes, evolution is taught as a science because it is one. Down through the years since Darwin first proposed evolution, there have been arguments among scientists and changes of opinion as new findings were reported. The enormous body of information gathered from such varied disciplines as paleontology, biology, medicine, comparative anatomy, genetics, serology, and even man's own breeding experiments has solidified the fact that evolution did occur and is taking place now. The lifeblood of science is an attitude—one of gathering data, testing ideas, and changing them as the findings dictate. Evolution is a case in point.

On the other hand, creationism is a belief in how life came to be, based upon religious scriptures, which, since they are perceived by some to be the word of God, are not subject to change, testing, or debate. That is not science. Both creationists and evolutionists seek truth. The difference is in their methods. One is science, the other is not. Yet both points of view should be accessible to inquiring minds. That is education. The appropriate place for creationism to be discussed is in a cultural or religious context. To present creationism as science is no more logical than teaching oil painting in a physics class.

Q *How can one species of animal or plant develop into two separate species?*

A The original species may have members that become isolated from the rest of its kind. There could be a physical isolation into two subenvironments. For example, during a storm at sea, a few individuals of the same species are thrown into a lagoon by a large wave. They inbreed and adapt to the quieter, warm, and shallow waters of the lagoon. With time, they become so different from the original species that they would be unable to interbreed with them. Another way the same thing could happen is if one group stopped interbreeding with the rest through sexual preference. Thus, the genes would not mix even though all members of the species remain in the same environment. This "genetic drift" would ultimately result in a new species. When two groups can't interbreed, they are classified as separate species.

Q *Why do scientists say we evolved from a monkey?*

A What is interesting about your question is that this idea was doubtless voiced by someone opposed to the concept of evolution who was stating what he or she thought scientists were saying. Scientists hold nothing of the kind. Man did not evolve from the monkey and may not even have evolved from the ape. Certainly we are more closely related to a monkey than, say, a chicken. But at best we are only cousins.

Scientists assert that man, monkeys, and apes evolved from a common ancestor several million years ago. The data supporting this conclusion are not as skimpy and speculative as some make out. But scientists are notorious for playing down the information that they have. This is why even today scientists refer to evolution as "theory" when in reality that evolution has occurred is hard fact, based upon many lines of evidence.

Q *I am amazed at the delicate balance between living things and the environment. For instance, if the planet's temperature was hotter or colder, so much life would be ended. Has science an explanation?*

A We think so. Consider the offspring of a particular species; some individuals will not be able to cope with the environment, for example the food supply or climate. These individuals will die. The hardier individuals will survive and pass on their traits to their offspring. Thus, a species will become, perhaps over thousands of years, better and better specialized in dealing with the environment in which it must live. Scientists consider this process—a "gene pool shift"—important in the development of a new species, one that is well adapted to all aspects of the environment. There is a danger, though, in that should a radical change in the environment occur quickly, a specialized species may not have time enough to adapt, and this could lead to extinction.

Q *Is the giraffe's long neck the result of stretching for leaves on high trees?*

A No. In the last century, some scientists such as Lamarck thought that acquired characteristics of an individual could be passed on to its offspring. Likewise, traits not used much would gradually disappear in the offspring. This was called the law of use and disuse, and is no longer held today. If it were true, then a weightlifter (for example) would pass on his built-up muscles to his children, which we know doesn't happen. Traits such as the giraffe's long neck are inherited via the genes. Where genetic characteristics represent an environmental advantage to the organism, the latter will survive and pass on these successful traits. Traits that do not equip an organism for survival within its environment will doom it to extinction.

Q *Humans walk upright while other primates don't. Why is this said to be a big "evolutionary advance"?*

A Try getting down on all fours and driving a nail with a hammer. You can't. If the forelimbs and hands are involved in support, this will severely restrict their use. However, standing on two legs frees the hands and arms for other tasks such as toolmaking, especially when the hands are equipped with opposable thumbs. It is this ability to make tools (and ultimately tools to make other tools) that accelerated man's evolutionary process to its present degree. Another point is that in standing, the senses of sight and hearing conferred a big advantage in surveying the surroundings, a big plus for survival. If you add the effects of upright posture and freed hands, the result is a creature capable of making not only tools, but also cave paintings, telescopes, and computers.

Q *How do insects become immune to insecticides?*

A It is not that the individual insect becomes immune. Most of the insects will die right away, but a few will have a built-in immunity, will survive and produce offspring that inherit this immunity to the particular insecticide to which they were exposed. This type of reaction is common to all living organisms and is a fine example of evolution in process. Another example is the current invasion of New York City by "superrats." A few years ago the city made a determined effort to poison all the rats in town, and at first the number of rats declined dramatically. But there were a few that withstood the poison. They have now repopulated their kind and the problem is no better. Fortunately, there are still no varmints who are immune to a good stamp with the foot.

Q *Isn't the "population explosion" just a myth?*

A Yes and no. The idea goes back to Thomas Malthus, an economist who lived from 1766 to 1834. He said that population increases geometrically while food supply increases arithmetically, meaning that the population would outstrip the food supply and people would die of starvation. To illustrate the meaning of this, start with one person and double the number each year for 30 years and you will have more than one billion people. In industrialized societies such as the United States, this has not held true because economic output has tended to lead population growth, so nobody starves. This is also due to widespread birth control. On the contrary, in nonindustrialized countries—the Third World—Malthus's theory holds true and people are indeed starving because the population growth is unchecked. If the human race cannot find a balance between food supply and population growth, the Four Horsemen—death, disease, war, and pestilence—will ride again.

Q *Why will the population explosion cause war, disease, and the like?*

A It is a well-established biological fact that animals such as birds, dogs, and cats will "stake out" a territory that can't be invaded without a fight. As long as territories are broad, no fights will occur. However, compression of territory by multitudes of individuals leads to quarrels. It will be instructive to consider a biological experiment in which white rats were provided with adequate space and food and allowed to reproduce with no restriction. At first, there were no problems—all rats had food and space. However, once the laboratory facility became overcrowded, fights resulted, and rat killed rat. In such conditions, disease is more communicable. This may be the reason why there is more crime in overcrowded ghetto areas. We really ought to spread ourselves out. About 98% of the population of the United States is concentrated in only 2% of the land area. In other words, most of our country is empty. A better distribution would be helpful for all.

Q *What is cloning?*

A Cloning is nothing more than a method of reproduction. If a human couple produces a child, the child would not be a clone because there has been an interchange of genetic material between the parents. If a simple cell divides into two cells (asexual reproduction), that would be cloning because the genetic material is identical and without contribution of genes from outside. Experiments have been attempted since the 1950s to produce replicas or clones of lower animals, with some success. Whether or not human clones can be produced is a moot question. Right now, adult cells cannot be cloned, so the closest duplicate of yourself you will see is probably in a mirror. If technology permits duplication of humans some day, it will raise a number of very interesting moral and legal questions.

Q *I still don't understand cloning. Of what use is it?*

A It is a process of duplication. As you may know, in nature there are organisms that reproduce asexually; in other words, an individual produces one or more other individuals without combining or exchanging genetic material. Thus, the offspring are carbon copies. In higher organisms such as man, there are always unique combinations of genetic material so that each of us is different in some ways. Research in cloning can be beneficial to mankind in providing a source of insulin, growth hormones, or other useful substances. Advances can also be made toward the cure of disease. The science fiction notion that you might walk down the street and bump into your exact double is a bit far-out.

Q *Did Charles Darwin believe in God? His idea of evolution opposes so much religious thought.*

A Darwin died a little before our time, but our reading of his life and work suggests strongly that he was a shy, modest, and deeply religious man. In fact, at one point in his life he was studying to be a clergyman. His life's work, leading to the concept of evolution in his role of biologist–geologist, actually produced within him considerable confusion about design in the universe, but we have no evidence he ever abandoned the idea of a Creator. Perhaps toward the end of his life he could be described as an agnostic. In some religions, of course, evolution is regarded as anti-God. But not all. The Catholic Church accepts evolution as God's method of creation, the concept being that man's body evolved as science claims but that at a certain point a soul was instilled into man apart from evolution. Contrary to some opinions, many scientists who espouse and even teach organic evolution believe in a Supreme Being.

Q *Is it possible to think without using words?*

A We suppose anyone can form mental pictures of various objects such as house and car, and you can even see yourself pulling into the driveway and walking into the house without thinking the words, "I am going home." We will think at times in such pictures, perhaps combined with words. However, there is a problem when it comes to abstract ideas and concepts. What mental picture comes to mind with the idea of truth? We believe that the most effective and far-ranging thinking is done when an individual knows a lot of words. Words are efficient symbols, and the more such symbols you can place into numerous combinations, the more creative potential for your thought. Surveys we have seen indicate that the most successful people in our society have the largest vocabularies.

Animals

Q *Are stingrays fish? Are they dangerous to humans?*

A There are quite a variety of stingrays, or "rays," in general belonging to the same group (the Chondrichthyes) as the shark. This means that they have cartilage rather than bone as the major part of their skeleton. They are fish. Some species are able to deliver an electric shock while other species have a sharp, barbed tail that is poisonous. We would assume that if you wish to pursue these creatures on the sea bottom, it could be dangerous as they will naturally defend themselves. Yet they are fished for on a commercial basis and the meat is very good. The larger species (e.g., the devil rays) are 20 feet across and weigh up to 500 pounds. Usually they lie on the seabed, well camouflaged and waiting for food such as other fish and crustaceans. When they swim, they resemble large bats flapping along.

Q *I saw a film on the piranha or cannibal fish. Is it true that piranha can reduce a cow to a skeleton in a matter of seconds?*

A This is perhaps a bit exaggerated, but the cow involved may argue the point. Schools of these ferocious fish, found in South America, have been known to attack a 100-pound capybara (a large South American rodent) and reduce it to a skeleton in one minute. Piranhas have also attacked humans and although we cannot cite specific cases, probably with fatal results.

Q *Can an electric eel electrocute a person?*

A Actually, the electric eel is not an eel, but rather a freshwater fish more closely related to the knifefish of aquarium hobbyists. Electric eels can exceed a length of 3 feet, and some reach up to 9 feet in length with a thickness of 8–10 inches. The body of the fish constitutes an electric storage battery in which the head and tail are negative and positive poles which, in contact with the victim, deliver a shock. The wider the spacing between head and tail in contact with the victim, the greater the shock. A charge of 1000 watts has been recorded for brief contacts. Such magnitudes are sufficient to kill even as large an animal as a horse. Humans have been attacked and severely stunned, but we know of no deaths.

Q *Is a barracuda as dangerous as a shark?*

A Some attacks on swimmers that were thought to be due to sharks were in fact attacks by barracuda. However, it should be remembered that the normal diet of the barracuda is other fish and not people. A barracuda can grow to ten feet in length and is extremely swift. They have razor-sharp teeth and attack anything that moves, which perhaps accounts for their unpredictable behavior. Some barracuda actually control herds of other fish in shallow water, waiting until they become hungry to eat them. Fishermen like to fish for barracuda because they are hooked easily, fight like tigers, and are a tasty item on the menu.

Q *Can sharks smell human blood in the water from a mile away?*

A You would have to assume that traces of the blood could circulate to distances of a mile in order for the shark to have the opportunity to detect it. This is highly unlikely unless you are bleeding like a stuck pig. It is known that people have been attacked by sharks in shallow water and have suffered severe wounds (and loss of blood) without the shark resuming the attack. In some experiments, sharks swam through human blood without getting excited at all. The same with rat blood. However, when dead rats were soaked with fish blood, they were immediately attacked by the sharks. Humans are not a normal part of the shark's diet and so we do not command particular attention. Actually, sharks are more attracted to vibrations and movement in the water near the surface where disabled or dying fish might be located. Perhaps this accounts for many shark attacks on humans. It may be a case of mistaken identity.

Q *Do fish sleep? They seem to be moving around all the time.*

A It isn't known if they sleep in the same sense that we do, but they do seem to have regular rest periods. Fish may appear not to sleep because, lacking eyelids, their eyes are always open. Some fish merely remain more or less motionless in the water, while others rest directly on the bottom, even turning over on their side. Some species are observed to dig or burrow into bottom sediments to make a sort of "bed." Some fish seem to prefer privacy when they rest, because their schools disperse at night to rest and then reassemble in the morning.

Q *Do ostriches really hide their heads in the sand when frightened?*

A Not unless it is an extremely cowardly ostrich and even then we doubt it. More probable are faulty observation and unwarranted conclusions about the behavior of ostriches. The ostrich must arc its long slender neck to reach food on the ground or in bushes, and this led to the notion that the bird was trying to stick its head into the sand. In addition, the ostrich often sits on the ground with legs folded and its neck and head stretched out along the ground, observing the surroundings with its keen eyesight. To a distant watcher, only the body was visible, and the head and neck were thought to be buried. The ostrich is actually a noteworthy bird. It is the largest living bird, the only one with two toes, and it lays a whopping three-pound egg. If frightened, the ostrich need not bury its head in the sand because it is perfectly capable of leaving any unpleasant scene at a graceful 40 mph.

Q *Hasn't Hollywood exaggerated the danger of big snakes, tigers, spiders, and other jungle creatures?*

A From our experience, we would agree with you. There is the tendency to magnify the danger presented by these creatures for the sake of drama and excitement. Most of nature's creatures simply want to survive and mind their own business without fighting with others. We have walked within five or six feet of boa constrictors without so much as a ho hum. Lions will lounge within a short distance of zebra herds without attacking because the lions are simply not hungry. Disturb a spider web, and the spider usually flees. While these creatures may present a menace under certain circumstances, they are not what you would call "mean." Man's greatest enemy in the jungle is perhaps the mosquito or other insects, while at home the worst enemy might be ants in the kitchen. In any case, we would feel safer in the densest jungle than in New York's Times Square at rush hour.

Q *Why have alligators and crocodiles been able to survive? They don't seem to pay much attention to their young.*

A There is a bit of a mystery here, it is true, because the mother is an indifferent parent. After laying the eggs (from 20 to 70 of them), she guards the nest only on a part-time basis. It is known that small mammals such as raccoons will eat the eggs of the American alligator. At the point of hatching, the female will assist the little gators into the water. Perhaps the major reason the crocodilia family has survived is that, once hatched, the young are perfectly able to take care of themselves without nurturing. The greatest enemy of these beasts is man, who hunts the adults for their hides to make luggage. This alone has brought some species to near extinction. Despite the gaping mouth and numerous teeth, they can't chew their food, being forced to swallow it in big chunks. The crocodilia are closely related to the dinosaurs as both groups belong to the subclass Archosauria. Somehow, they escaped the fate of the dinosaurs, but for how long?

Q *Why do snakes flick their tongues in and out?*

A A snake's tongue is an important part of its sensory organs, a sort of combination of taste and smell. The tongue may pick up a particle from the air and transfer it to two depressions in the roof of the mouth where its meaning is further evaluated.

Snakes usually have good eyesight as well, but are deaf in our understanding of the word. They can sense vibrations passing through the ground. Although snakes are often considered a symbol of evil or the Devil, as in the biblical account of Adam and Eve, they have their good points, a major one being that they eat rats and other rodents. This saves millions of dollars in agricultural products each year, and cuts down on the spread of diseases carried by rodents.

Q *An old farmer told me that rattlesnakes are deaf and can climb trees. Aren't these superstitions?*

A The farmer knew his stuff. While rattlers are deaf, they have great sensitivity to vibrations. One reason you may not see many such snakes in the woods is because they sense vibrations from footsteps and get out of your way before you see them. Yes, rattlers can climb trees, and also swim, but they don't do this too often.

Q *I have never seen an answer to this age-old question: Which came first, the chicken or the egg?*

A You could also ask: which came first, the turtle or the egg, or the alligator or the egg, or any of the creatures that lay eggs. We don't think the answer that tricky if you take into account evolution. All creatures on the earth, past and present, represent life forms that did not previously exist, but yet had to have evolved from previously existing types. In the case of the chicken, the ancestors evolved over millions of years, with successive generations becoming more and more chickenlike. Finally, the most chickenlike of the chickenlike predecessors hatched an egg. It was a true chicken. So the obvious answer is that the egg came first, but it required a very close relative to oversee the nativity.

Q *Do any birds have teeth?*

A There are no birds living today that have teeth, although some have toothlike projections or notches along the mandible, but these are not true teeth. Actually, the first birds did have teeth and were closely related to the dinosaurs. The first bird was about the size of a crow and had a full set of teeth. It was found in southern Germany under such excellent conditions of preservation that the feathers could be seen. Otherwise, the remains would have been classified as a reptile. The name given to this creature was *Archaeopteryx.* From that beginning about 150 million years ago, birds became widespread and grew larger, and most of them had teeth, including a few bigger than the ostrich. One species laid an egg fully 13 inches long. The success of birds is underscored by the fact that there are 25 birds for every person on this planet, or more than 100 billion birds. Despite this, there are some birds threatened with extinction, such as the whooping crane and the California condor. The passenger pigeon used to darken the skies but is now extinct.

Q *Is it true that the "pelican can store in his beak enough food for a week"?*

A What is thought of as a storage place in the lower beak is an elastic pouch that serves mostly as a fish-capturing device rather than a storage place. The pelican is nonetheless a remarkable bird. The pouch also serves as a cooling mechanism by providing more surface area to remove heat, particularly in the young. The pelican may be clumsy walking on the ground, but is a superb flyer, and takes long plunging dives from great heights to catch fish swimming close to the surface. Oddly enough, the pelican is a silent bird, without a song or a screech. One species of pelican is very clever. It bands together and "herds" fish into shallow water against the shoreline so they may be caught by their colleagues waiting there. To complete the limerick you quoted in your question, "I'll be darned if I know how the helican."

Q *I sometimes hear the expression dead as a dodo. Just what is a dodo?*

A It is an extinct bird that once inhabited the islands of Mauritius and Réunion in the Indian Ocean. It is another of those creatures whose contact with man led to its extinction. When settlers arrived on those islands, they brought pigs, which apparently feasted on the eggs of the dodo. This led to their demise, sometime during the 17th century. The dodo was about the size of a turkey, could not fly, and was somewhat clumsy. Although living specimens were sent to Europe, there is today no complete bird on display in any museum that we know of. The dodo was a distant relative of the pigeon.

Q *Are there more canaries in the Canary Islands than in any other place in the world?*

A There is a substantial canary population in the Canary Islands, but there are also more than 200 species of other birds as well. The yellow canary we are familiar with actually was derived from native green and brownish canaries found on the islands by 16th century European explorers. However, the Canary Islands were not named for the canary populations. About 2000 years ago, African explorers had reached the Canaries and were impressed by the large dog population. When the Romans heard about this, they began referring to these islands as Canaria, from the Latin word for dog—*Canae*. The Spaniards eventually took control of the Canary Islands and still call them, in Spanish, Islas Canarias. We might add that the Canaries are of volcanic origin, with mountain peaks rising to more than 10,000 feet. Despite its tropical appearance, with oranges and bananas growing at lower elevations, the mountain peaks often are covered with snow.

Q *Do army ants actually eat people?*

A No. Although, army ants may ingest larger carcasses, they will not attack and eat a living person, despite what Hollywood may depict. Instead, army ants feed upon cockroaches, beetles, locusts, spiders, and other insects. And they are vicious. Any of these unfortunate insect victims would be covered instantly by a dense mass of ants and their bodies torn apart. In seconds, limbs and other pieces of the insect are severed by sharp pincers and moved to the rear of the ant army.

Usually army ants scour a forest, especially tropical forests, in columns. Upon finding a target, the column divides into two directions to cut off escape. The ants resemble a real army, even foraging with advance scouts ahead of the column, attacking an area and abandoning it as its resources are devoured. Army ants will not attack other ant species except for the cowardly Hypoclinea, and even then only the larvae and pupae.

Q *Is it true that ants take slaves?*

A Yes. There are five ant species that do so. The belligerent queen of the raiding group and her consorts force their way into another ant group's nest. If any worker ants resist the invaders, the queen quickly kills them. There are two ways in which the vanquished ants become slaves. (1) The discombobulated ants stop fighting the victorious queen and attempt to guard the brood of larvae. When the larvae grow into ants, they recognize the foreign queen as their own. The original queen will either have died of neglect or been killed during the invasion. (2) The invading ants gather up the pupae and larvae and take them back to their own nest where they develop and work for the other ant group.

Q *How can beekeepers handle hives with their bare hands and not get stung?*

A Sometimes they do get stung, but some experienced beekeepers prefer to work without protective gloves. They do wear a veil over their heads and shoulders and generally employ a smoking device with a bellows to quiet the bees. Beekeepers have also learned that slow deliberate movements will not excite the bees. Also, bees chosen to inhabit the man-made hives are those that produce the most honey and are the most docile. A single hive may produce 30 to 40 pounds of honey in a season. Beekeepers have more to worry about than getting stung. They have to guard against predators such as skunks, toads, and mice that invade the hives not only for the honey, but to eat the bees as well.

Q *I have found these funny bugs in my closet which a friend says are silverfish. Can you tell me more about them?*

A They are about one-half inch long with three bristles for a tail, and a silvery appearance; they are named for the small, silvery scales they possess. They are not dangerous in that they bite, but they feast on anything with starch, including wallpaper (the glue), and can also damage books. Silverfish are indoor insects and are found all over the world. They live for as long as two years, but if you don't want them for roommates for that long, any good insecticide should do the trick.

Q *I saw a beetle that was about two inches long with a terrible-looking horn. Are these dangerous?*

A This sounds like the rhinoceros beetle but it is difficult to be certain because there are 25,000 kinds of beetles in the United States alone and about 250,000 kinds of beetles worldwide. They are the largest group of animals known. Some of them are certainly dangerous. In Spain, one variety causes skin blisters upon contact. In Sri Lanka, they are parasites upon the internal organs of small children. They can also carry germs causing diseases, such as the Dutch elm disease that has killed so many trees. Still, some of them can be beneficial. The European ground beetle was introduced into the United States to control the injurious gypsy moths. Then, of course, there were the four beetles from England that rocked the whole world and whose buzzing can still be heard on radio.

Q *Why do I have so many fruit flies? They seem to come out of nowhere.*

A What we tend to call a fruit fly isn't a true fruit fly at all but rather a vinegar, or pomace, fly with the Latin name *Drosophila*. They do, however, lay a large number of eggs on fermenting or ripe fruit, and the parents can produce thousands of offspring in a short time—short, because these little flies have a life cycle of only two weeks. *Drosophila* and the true fruit flies have caused considerable damage to citrus crops, especially the Mediterranean fruit fly. Nonetheless, *Drosophila* is very useful to scientists because of its short life cycle and large chromosomes, which are easy to see and study. From study of these flies, we have gained considerable knowledge of genetics and the workings of evolution.

Q *I had a lot of trouble getting rid of the fleas on my cat, and I wondered if fleas serve any useful purpose in nature.*

A Fleas are parasites that feed on the blood of animals and in the process carry and transmit any number of diseases from one host to another. One such disease was the bubonic plague that struck Europe during the Middle Ages and was spread by rats infested with fleas. You have to dig deep to find any redeeming qualities in the flea, and the best we found was the deliberate introduction of fleas carrying a disease into the rabbit population of Australia. The disease involved is myxomatosis, and it kills rabbits. At the time, Australia was overrun with rabbits, and the disease-carrying fleas were used to control them. Fleas are also useful in flea circuses to display their jumping ability and strength, if this can be seen as useful. A flea has powerful hind legs and can jump ten inches straight up. This is equivalent to a man jumping 300 feet into the air. This is perhaps the only thing notable about the flea.

Q *How long do spiders live and are they all poisonous?*

A Most spiders seem to live about one year, but they can live longer in warmer climates. The large bird spider lives longer than most, and one in captivity lived for 20 years. Generally, as for humans, the female lives longer than the male. Yes, almost all spiders have some kind of poison.

It is enough that the poison paralyze or kill the prey, which are quite small compared to people. The only really dangerous spiders are the black widow and the brown recluse and even their bite is only rarely fatal to man. Spiders have an undeserved reputation as being evil; they are helpful in eating many insects.

Q *Is silk made only by silkworms in China?*

A No, but it used to be. The mulberry silkworm was known only in China and the secret of making silk from their cocoons was a closely guarded one. In fact, the Chinese would torture to death anyone caught teaching foreigners how to make silk. In this way they were able to keep their secret for nearly 3000 years. Finally, mulberry seeds and silkworm eggs were smuggled out of China and by the Middle Ages, the art of silk-making had spread to other parts of the world. To produce the raw silk, the worm secretes two substances from two separate glands. As these two substances are extruded, they come together to form the silk threads, similar to using two substances to form epoxy glue (another case of nature inventing something before man). Other insects, of course, produce a kind of silk, for example, the spider's silk. But this is too sticky for commercial silk. Despite all the synthetic fabrics now available, silk is still much prized around the world and is considered the "royal cloth."

Q *Scientists can train white mice to run through mazes. Have they been able to do that with lower creatures such as insects?*

A There is growing evidence that insects can learn from their mistakes, contrary to the general belief that insects are instinctive and stupid. In recent experiments at the University of Illinois, praying mantises were fed a diet of milkweed bugs, which in turn had been fed either of two diets. One of these diets was a toxic substance and apparently tasted bad to the mantis, and so was rejected by them. Subsequently, they also refused the other group of milkweed bugs that had been raised on nontoxic sunflower seeds. The mantises also refused any insect disguised to look like a milkweed bug. There are any number of other examples showing that insects may have more smarts than we credit them. Those who claim insects will take over the world if man becomes extinct may not be far off the mark.

Q *Do anteaters eat only ants?*

A As a matter of fact, they do—that is, ants and termites. If a supply of ants and termites is not immediately available and the anteater is hungry, it will eat whatever insects are present. In the same way, if you do not like spinach, but are starving, and only spinach is available, you will eat spinach. In this respect, animals are as logical as we. There are several varieties of anteaters in both South America and Africa. They have sharp claws for digging out anthills and a long sticky tongue for lapping up the unfortunate ants. Some anteaters live in trees, sleep by day, and prowl for anthills by night. In Africa, the anteater is the aardvark. It is not an aggressive creature and minds its own business. When it is attacked, however, its sharp claws are a good defense, even against lions. Probably the greatest threat to the anteater is man, because some anteaters have teeth that are thought to be magic charms against evil. Anteaters in Central and South America have no teeth, but the aardvark does.

Q *Can a bat really get tangled in my hair?*

A This is a myth. With its echo-sounding system, a bat can easily avoid not only your hair, but the rest of you as well. Perhaps the basis for this myth is that when a bat hibernates, it becomes sluggish and not in full control of its flying ability. Under such circumstances, a disturbed bat may on occasion have struck someone in the head. In our Western culture, we associate the bat with graveyards, darkness, and things evil. Demons have bat wings, not angels. Yet, oddly enough, the Chinese regard bats as symbols of happiness and longevity. This is logical because for a small mammal, bats have long lives—some have lived up to 21 years. Another thing about bats that isn't generally known is that their homing instinct rivals that of the homing pigeon. The little brown bat has been banded and shown to fly 60 miles to its home in a single night. And it can do it blindfolded!

Q *How long can bears hibernate without eating?*

A While there are numerous insects, amphibians, and reptiles as well as mammals that hibernate, oddly enough the bear is not a true hibernator. It does gain fat and, when winter arrives, sleeps for long periods, but not continuously. At irregular intervals, it arouses and wanders about, but doesn't eat much. In fact, by spring, its intestinal track is in a state of partial collapse. Some small mammals such as the marmot are true hibernators and may dig burrows as deep as 30 feet underground to snooze away the season. It should be noted that some creatures sleep during the summer rather than the winter; this is called estivation. For example, during drought, some lungfish will burrow into the dried-up mud of river or lake bottoms and wait for more water to come. The champion hibernators must be those bacteria that have been aroused from dormancy after several million years.

Q *Hyenas and jackals are usually held in low regard as cowardly beasts. Do they deserve this reputation?*

A Not really. Both groups are timid and retiring in many situations, but can be formidable and dangerous if they are hungry or their offspring are threatened. A major reason for the poor reputation of these animals lies in their nocturnal habits and their diet of the remains of kills of other creatures such as lions. They often follow the lion, eating what the lion leaves behind. Owing to this propensity to eat dead flesh in the darkness, these animals have come to be associated with evil and the Devil. In addition, the laughing bark of the hyena can often sound human, and so stories have arisen of men changing into hyenas for evil purposes. However, hyenas have sometimes been known to lead lost persons safely out of the wilderness.

Q *Are there any authentic cases of packs of wolves killing and eating humans?*

A We know of none. Human flesh is not a normal part of the wolf's menu, although they have voracious appetites. The notion of a pack of 20 or more wolves chasing sleighs or surrounding the trapper's cabin in a snow-swept forest is pure fiction. The timber wolf is the common species and they keep to themselves in groups of five or six. True, they have a fondness for sheep and other domestic animals, but their main diet consists of rodents and rabbits. On occasion, they are quite capable of chasing down larger prey such as deer, elk, or horse. The idea that wolves eat humans may have derived from the sight of wolves sniffing around corpses on battlefields in earlier times when wolves were abundant. If anything, the wolf has a beef against us. Man has exterminated the wolf in the British Isles and western Europe, and nearly so in North America. While the wolf is most often viewed as a symbol of evil, there are many stories wherein the wolf is friendly.

Q *How long can a camel go without water? Are they really stupid animals?*

A We haven't conducted any IQ tests on camels lately, but they are truly remarkable animals. A camel can go for 17 days without drinking any water. Carrying a load of 500 pounds, they can travel 75 miles over a three-day period, again without drinking any water. There is a secret to this. The camel carries a great deal of fat in its hump and has the ability to manufacture water out of this fat by oxidation. This is not to say the camel doesn't get thirsty. When it gets the chance to drink after a long drought, it can suck down 25 gallons of water. Camels can live 40 years. Oddly enough, the camel originated in North America, where it no longer exists. We might add that camels have very pretty eyelashes.

Q *According to a recent television commercial, there are 12,000 wild horses out West. I didn't think wild horses existed anymore.*

A The commercial's estimate is too conservative. According to our information, the wild horse population on public lands in the West is now surpassing 45,000 and growing because the horses are protected and there are diminishing numbers of natural predators. If these horses are to be truly protected, we must be concerned about the carrying capacity of the land. If the food supply runs out, then these horses will starve to death. The land is already being overgrazed. We are reminded of the small herd of reindeer introduced onto St. Matthew Island in the Bering Sea during the 1940s. The island had an abundant food supply and no predators. The population swelled to 6000. They faced a severe winter in 1963 in a starving condition, and were wiped out. There just wasn't enough food to go around.

Q *Why are foxes regarded as especially clever?*

A Members of the dog family, which includes foxes, are intelligent. In addition, foxes have above-average hearing, sight, and smell. These advantages can make any animal look good, as when being pursued by hunting dogs. There are a variety of foxes, but the common ones are the red fox and the gray fox. The red fox seems to be smarter, but the gray fox knows how to climb trees. They are generally shy, retiring animals that live mainly on rodents and rabbits, but will take a farmer's chicken if given the chance. Actually, the fox's chief enemy is man, as the fur is desirable for such varieties as the silver fox. Yet the fox helps man by eating destructive rats and mice. Foxes can be domesticated.

Q *Why do cats purr?*

A In our experience, cats that are purring seem to be pleased with themselves, but they are also known to purr when in extreme pain. Most authorities suggest that purring, as a vibratory signal, is a homing device for kittens to come to the mother for milk. Not all members of the cat family purr. Lions and other big cats roar instead.

There is a widespread notion, while we are at it, that cats hate water or cannot swim. Many cats indeed like water and swim quite well. What they don't like is being suddenly doused with cold water, an attitude not unlike ours.

Additionally, one member of the cat family, the cheetah, is the fastest mammal alive. It can attain short bursts of speed of up to 70 mph.

Q *What is one year in a cat's life equivalent to in human years?*

A It depends to a large extent on the cat's situation. If the cat is left to shift for itself in the outdoors, and becomes feral, it may live to about age eight. Compared to a human living to age 72, a one-year-old cat would be nine in human terms. On the other hand, domesticated cats that enjoy a regular balanced diet and the protection of a fond owner often live to age 15 or even older. In the latter case, an eight-year-old cat, instead of being at death's door, would be comparable to a human of age 40–45. To some extent, this indicates the importance of proper care to extend the life span of any animal, man included. The situation with dogs is similar. A dog reaches middle age (compared to us) at about age six. Many dogs live ten years and some into their teens.

Q *Are there actually white elephants, and if so, why are they thought to be bad luck?*

A Yes, there are white elephants, although they are rare albinos found mostly in Thailand and Burma. In those countries, they are considered sacred and often are raised by human nurses to ensure their survival. They are not truly white but rather a grayish white with pinkish eyes. The great showman of the last century, P. T. Barnum, in his efforts to bring to the public the most unusual of creatures for his famous circus, obtained at great cost a white elephant to exhibit. Such an elephant is expensive to maintain, for it eats up to 500 pounds of food a day and drinks perhaps 60 gallons of water. Unfortunately for Barnum, the white elephant was disappointing as an attraction to the public. Barnum not only lost money, but was unable to find anyone who would buy the animal. Hence, the expression came into our vocabulary of having a "white elephant," something expensive to keep and tough to get rid of. We have not found out how Barnum finally got rid of the elephant.

Q *Why can't I detect the blind spots in my eyes?*

A Each of our eyes has a circular blind spot lying within our overall field of vision, but the left eye "covers" some of the blind area of the right eye and vice versa. Also, our eyes are moving almost constantly. Thus, even a one-eyed person would probably not notice the blind spot. The reason for the blind spot is as follows: when light enters the eye, it impinges on the retina and excites light-sensitive cells. The images are conveyed through optic fibers to a gathering point where the fibers bundle together to form the optic nerve, which then exits as a sort of "cable" carrying the visual signals to the brain. The place of entry of the optic nerve into the retina has no light-sensitive cells, and is the blind spot.

Q *Is it true that the human body can secrete its own morphinelike substance under certain conditions?*

A Yes. The substance is called endorphin and is known to alleviate pain and promote a state of euphoria in an individual under great stress. For example, consider the following report: a hunter was attacked and seized by a lion, which then started to drag him into the jungle. At first, the hunter was frightened but this feeling changed to indifference, and, although realizing he was going to die, he did not mind at all. We know about this because the hunter was rescued. It has been suggested that long-distance runners also secrete endorphins, accounting for the well-known "runner's high." It is suspected that witch doctors edge into apathetic trances by endorphin secretion resulting from continuous dancing.

Plants

Q *Can plants grow from seeds hundreds of years old?*

A Yes, although the longevity of seeds varies widely from one plant type to another. Some seeds may last only days or weeks before losing the power to germinate while others are good for years. In fact, seed of the Oriental lotus—found in a peat bog in Manchuria and dated at about 1000 years—produced flowers. It is of importance that seeds be distributed some distance before taking root. The wind is an important distributing agent. Grass seed has been found at heights of 3000 feet, and of course everyone is familiar with the little parachutes of the dandelion. Seeds also spread via water, such as floating coconuts, and other seeds are carried to new areas by animals, including birds.

Q *Why are some mushrooms called toadstools?*

A The term "toadstools" refers to mushrooms that are inedible or poisonous. When a toad is alarmed or attacked, the warts on its back secrete bufonenin, a poisonous substance if taken internally. Curiously, this is the same chemical produced by the warts of the poisonous mushroom *A. muscaria*. Perhaps that is the connection. Mushrooms have been eaten for thousands of years and were considered a delicacy by the Greeks and

Romans. Ancient priests were so possessive of mushrooms that they forbade ordinary people from eating them. Although tasty, mushrooms have little food value and may consist of 90% water. There are many varieties of edible mushrooms but also a number of poisonous ones that can actually be fatal. One who collects and eats mushrooms indiscriminately is asking for trouble. Some mushrooms are hallucinatory.

Q *Where does cork, such as that used as bottle stoppers, come from?*

A Cork is the bark from a type of oak tree found in the Mediterranean area, especially in Spain, Portugal, and North Africa. Cork is filled with air bubbles that provide insulation for the oak tree against the changing weather. These trees don't produce much cork until they are more than 50 years old. When the cork is cut, say, to make stoppers for bottles, the air bubbles are also cut to make little suction cups that cause them to hold tight. Because cork has only one-fifth the density of water, it is quite as useful in life preservers as in bottle stoppers. It is also useful for insulation and soundproofing.

Q *I love vinegar on salads and cucumbers. Just how is vinegar made?*

A This was probably discovered accidentally thousands of years ago, along with the discovery of how to make alcoholic beverages. If you allow something with sugar in it (such as a fruit, e.g., grapes) to ferment with yeast, you get wine. If you allow this to react with certain bacteria, a very weak acetic acid is produced, and that is vinegar (the word means "sour wine"). So vinegar is basically acetic acid. It has many uses besides on salad. It is a preservative of course, but also has medicinal value, having been prescribed since the times of Hippocrates, the Greek physician. It was used during the Civil War to curb scurvy. It can also take the sting out of a nasty sunburn—if you can stand the smell.

Q *The three wise men brought gold, incense, and myrrh. What is myrrh? Is it valuable?*

A It is a gum resin exuded from the myrrh tree, which is a short, stubby tree 4 to 20 feet high found in Somalia and some Arabian countries. After the myrrh oozes out, it hardens as small irregular lumps. Many centuries ago, it was especially sought after in Europe for use as a perfume. However, it has a bitter taste; in fact, the Arabian word "murr" means bitter. It was also used in medicine as an antiseptic and in embalming. Its value today is slight. One novel use is in Christmas tree ornaments where a few small lumps of myrrh are enclosed inside clear plastic balls.

Q *Is the tomato a fruit or a vegetable?*

A Take your pick. A biologist would say it is a fruit, but in a Supreme Court decision (this sounds silly) in 1893, it was classified as a vegetable because it was typically served and eaten with other vegetables. So much for science. The tomato has an interesting history. It seems to have been a wild yellow species found in Bolivia and Peru, then cultivated in Mexico and shipped to Europe after Columbus. The Italians called it the Golden Apple because of its yellow color, but soon scarlet varieties emerged. In the United States, it seems to have been first grown by Thomas Jefferson in 1781, but a lot of people refused to eat it until as late as 1900 because it was believed by many to be poisonous. For many Europeans long ago, it was the "love apple" because it was thought to make the person more romantic. Tomatoes are of course neither poisonous nor romance-enhancers, but they are excellent sources of vitamins A and C.

Q *The flesh of a watermelon I cut into wasn't red but a bright orange. Was this a freak watermelon?*

A No it was not, and we'll bet you found it delicious, assuming you ate it. People are so used to the typical red-fleshed watermelon in the supermarket that they forget, or may not know, that there are several varieties ranging from shades of white to red to yellow. Actually, the watermelon is a member of the gourd family and is thus related to pumpkins, cucumbers, and squashes. Some watermelons may weigh 50 pounds. They have been grown for at least 4000 years, being depicted in murals from ancient Egypt. We hope that by some gene-splicing technique they can come up with a kind without so many seeds.

Q *Where are lemon trees found?*

A They are found throughout the Mediterranean area in Italy and north Africa and also the southwestern United States. These are mostly cultivated trees 15 to 25 feet high that produce rather incredible numbers of lemons. Some trees may produce up to 7000 lemons in a single year. The origin of the lemon tree is not known for certain. The early Crusaders found them in Palestine and brought them back to Europe. They are rare in China and

India. Italy is today a world leader in lemon production, but oddly enough, the ancient Romans didn't have any lemons. In some cultures, the lemon is thought to have magical properties. By driving several iron spikes through a lemon, you can ward off the evil eye. If you need to do this.

Q *There are so many products with lemon in them. Just what is it that gives lemon its flavor and smell?*

A There are oil glands in the lemon peel. The secreted oil contains organic compounds—aldehyde and ester. These compounds are responsible for the lemon's taste and smell. After the juice is extracted, it is concentrated by evaporation in ratios of 3:1 to 6:1. As you may know, lemons are rich in vitamin C and this is preserved by passing the freshly squeezed juice into a tank maintained as a vacuum; otherwise, the vitamin C would be destroyed by oxidation. One of the main uses of lemon is in making lemonade, and as might be expected, the hotter the summer, the more lemons are sold. We might speculate that the lemon's reputation as a "refresher" accounts for its popularity in such products as soap, perfume, and furniture polish. It seems the lemon is a latecomer in our civilization. Early Greeks and Romans did not use it, and it was only introduced into Europe about 1000 A.D. An average lemon tree produces about 1500 lemons a year.

Q *Who discovered how to make wine?*

A We do not know who, but there is a good chance it started in the Mediterranean or the Middle East, perhaps among the Greeks or Egyptians, or the early inhabitants of Italy. Its discovery may well have been accidental. We can visualize some grape juice "spoiling" and then being consumed. Thereafter, its preparation would have been deliberate.

Early wines must have tasted awful by our standards today. Wine was aged in goatskins stoppered with greasy rags. Air could still get in, which is undesirable. It wasn't until the 18th century, when bottles and corks came into wide use, that wine could be aged properly. Good or bad, wines were drunk thousands of years ago, as mentioned in the Old Testament, and we know the ancient Egyptians drank beer more than 5000 years ago during the pyramid-building epoch. Wine probably came before beer. The development of wine-making as an art was fostered by early monks and priests because wine was (and still is) used in Communion services.

Health and Nutrition

Q *The symbol of the physician is a stick with a couple of snakes on it. What does this have to do with healing?*

A Nothing. It is the symbol, two entwined snakes on a staff, of the messenger of the gods, Mercury. In ancient times, it was a badge of authority and proclaimed that its bearer was a sacred person who should pass unmolested. Thus, it can be interpreted as a sign that the physician was not to be obstructed in his or her errands of mercy.

On the other hand, we find that there is a Greek god of medicine called Asclepius, who probably was a real person. After his death he was deified, with festivals in his honor. Legends about him resulted in the establishment of a cult in ancient Rome during times of disease epidemics. Asclepius is always symbolized as a snake and he is depicted as carrying a club with one snake twirled about it. Physicians need to deduct one snake from their emblem if it is to have true medical relevance.

Q *On the old sailing ships, was scurvy a fatal disease?*

A In many cases yes, and thousands certainly died at sea. During and prior to the 18th century, an outbreak of scurvy was more dangerous to officers and crew than any maritime enemy. Symptoms included failing strength and severe mental depression followed by rupturing blood vessels with hemorrhage and loss of several teeth at a time. The connection between scurvy and diet was finally recognized. When crews were deprived of fresh fruits and vegetables for long periods, the disease appeared and took its toll. In 1795, the British Navy made a regulation that lemon juice be issued to all hands. This seemed to prevent scurvy. Later, limes were used, and thus the term "limeys" for British sailors. From that time on, scurvy essentially disappeared from ships at sea. It was not until 1932, however, that it was found that a deficiency of vitamin C was the actual cause. Although scurvy seems to have been associated with sailors, it was also prevalent among land populations whenever vitamin C-bearing food was in short supply.

Q *How can an earthquake start an outbreak of disease, such as was feared in Mexico City?*

A Because an earthquake causes the ground to vibrate quickly, some underground water lines and other utilities will be broken. Germs can thus enter the water supply. Further, such a disaster brings out rats and other carriers of vermin into closer contact with people. Bodies (animal as well as human) lying under rubble for many days, as was the case in Mexico City, are an additional danger for the spread of disease. And do not forget pets, such as dogs and cats, that may also be the victims of a quake. Epidemics, fires, and other secondary effects cause more casualties than the earthquake itself. A major consideration in the September 1985 earthquake was the fact that the city is built on a swamp and the event was aptly described as the ground jiggling like a bowl of Jell-O. Surprisingly, older buildings survived better than modern ones. Older buildings had thicker, stouter walls.

Q *Almost all countries seem to have the custom of toasting one's health with a drink. Why?*

A This is a very ancient custom, originally as an offering to the gods or to the dead to placate them and stay on their good side. Offerings of food and drink to the gods appear as a rite in many cultures. As time passed, it seemed a good idea to make a symbolic offering or acknowledgment to living people who are powerful or admired. To toast another's health is to say in effect that it is hoped they have long life. Lord Chesterfield once said, "Can there be anything in the world less relative to any other man's health than my drinking a glass of wine?" True, but it exemplifies how behavior based in some logic can evolve into something not so logical. Well, let's drink to that.

Q *Could it be that the Roman emperors suffered from lead poisoning?*

A We have taken a look at the evidence regarding this contention. The Romans did use lead pipes in their plumbing systems, and it is known that small amounts of lead would thus have entered their drinking water, foods cooked in water, and beer and wine. A few of the Roman emperors did display some apparent derangement but there are other illnesses with similar symptoms. And shouldn't all the emperors and many citizens of Rome have displayed signs of lead poisoning? Claudius was a gentle ruler compared to the murderous Caligula, yet they lived in the same palace and shared the same lead plumbing. In Roman times, as today, you did not need lead poisoning to be ruthless and evil, and life was cheap. We have no firm answer to your question, but it is an interesting speculation.

Q *When I see chocolate-covered ants for sale in the gourmet food section of supermarkets, I wonder how common eating insects is. Ugh.*

A When you consider the abundance and variety of insects—there are 250,000 species of beetles alone—it should not be surprising that they have been eaten in some cultures. Actually, insects are high in protein. Consider the following true story of a group of explorers who starved to death in northern Canada. Their bodies were found encircling a lantern around which were numerous (dead) moths that had been attracted by the light. Had the explorers eaten the moths, they probably would have survived. What is regarded as food, even a delicacy, to one culture, may be repellent to another. A lobster is certainly creepy looking, yet in our society it is a mouth-watering delicacy. If you offered a cold glass of milk to people in some South American countries, they would find it as repugnant as if you had offered an American a glass of fresh blood. Large white grubs found under logs are hardly tasty-looking to us, yet some Japanese consider them a delicacy. A hearty meal depends upon where you are and what is available.

Q *Could the Great Plague happen again?*

A We don't think so. The Great Plague, or Black Death, was one of the greatest disasters of all time. When it hit Europe during the 14th century, it was certainly the biggest catastrophe up to that time. It seems to have started in Asia and then spread to Europe and later England and Scotland. At least one-fourth of Europe's population were killed. At that time, it was not known that the disease was being carried and spread by rats and their fleas. The plague was at its height between 1349 and 1351, and continued at intervals to the end of the century. Relatively few cases of bubonic plague occur nowadays around the world.

Q *Why is sickle cell anemia found only among blacks?*

A It isn't found exclusively among blacks, but rather primarily. It is caused by a small change in the genetic code for hemoglobin production. In such cases, the red blood cells, when deoxygenated, acquire a sickle shape and are unable to carry oxygen. This leads to severe anemia. About 20% of the population in central Africa possess this trait.

There is an interesting correlation between sickle cell anemia and malaria. In malaria-infested parts of the world (such as central Africa), sickle cell anemia is prevalent and serves as a means of bodily resistance to malaria. Where malaria is not a big problem, the incidence of sickle cell anemia is greatly reduced. In other words, sickle cells act as a defense against malaria. Originally, sickle cell anemia was brought to colonial America by slaves from Africa. Because there is little or no malaria in the United States today, it is to be expected that sickle cell anemia will progressively decline because it is no longer advantageous. This would represent another example of natural selection, by which the American black population will benefit.

Q *Are there some people who experience no pain?*

A As far as is known, all people experience some degree of pain. It is nature's signal that the body is threatened. Without pain, we could very well damage our bodies without knowing it. For example, if you place your hand in a flame, the immediate pain causes you to withdraw the hand instantly. Otherwise, without pain, your hand might be burned beyond repair. A throbbing pain in the stomach area may be a signal of stomach cancer or perhaps an appendix problem, and to see a physician. Each of us has a level of pain we can tolerate. It seems that Eskimos can tolerate more pain than the average person. Also, a person can stand more pain at certain times than at others.

Q *A man named Thomas Parr is said to have died in 1635 at the age of 152. Could this be true?*

A It is highly improbable. A major problem is that throughout human history, most people were never registered at birth and, therefore, many claims of great longevity cannot be substantiated. The best documentation we found was for Pierre Joubert, who died in 1814, having lived 113 years and 124 days. It is likely that the human life span is genetically controlled and has not changed since ancient times. Thus, a Roman baby born 2000 years ago had the same potential as a baby born today even though the average life span was only 28 back then. We live longer today because of control of disease and other scientific advances as well as better hygiene and nutrition. It can be expected that more people will live to be 100 or longer.

Q *A man who recently died in Russia was supposed to have been 150 years old. Isn't that some kind of record?*

A It is very difficult to document such claims because birth records often were not kept that long ago. It is known that some men in Russia exaggerated their age to prevent being drafted into the army. In many societies, older persons are accorded great honor, so it may be to one's advantage to add on a few extra years now and then. In one case, a man came to be five years older than his mother. The oldest man on record is actually "only" 114 years and such longevity is extremely rare. However, there are regions in the world where the population as a whole is older, with many people living into the 80s and 90s. These areas—Ecuador, Pakistan, and parts of Russia—are inhabited by people of Indo-European stock who subsist on a carbohydrate, low-fat diet with very low caloric intake and engage in farming and rigorous physical labor. Meat is rarely eaten. Perhaps if we all followed this kind of life-style, we would all live longer.

Q *Why is green thought to be a restful color?*

A We have heard the theory that green produces in us a more secure feeling because our ancestors lived in forests or jungles, which of course are usually green. As far as we know, there is no physiological basis for this notion. Some people may equally find violet or even bright Chinese red to be ''restful.'' There is little question that color influences our lives in the things we buy, where we live, and the people we associate with. Even what we eat. In antiquity, colors were linked to certain traits: blue for loyalty, red for courage, yellow for love, and so on.

Q *I was reading about Arctic explorers and how they ate pemmican. What is this food?*

A Pemmican is a concentrated food first prepared by North American Indians. Pieces of deer or buffalo meat were cut into strips and dried in the sun. The strips were then pounded into a powder, seasonings added, and fat blended in. The word ''pemmican'' means ''long journey'' and the Indians used it on such journeys because it was nutritious and easy to carry and did not have to be cooked. European explorers, quick to see the advantages, started using it. However, with the advent of modern methods of freeze-drying and other ways of preservation, the use of pemmican declined early in this century.

Q *Why do some people refuse to eat oysters in months lacking the letter ''r''?*

A The belief that oysters should not be eaten in May, June, July, or August has an ecological basis. Certain poisonous microorganisms are present in high levels during these months. Oysters and other shellfish that feed on these microorganisms can be contaminated with toxins harmful to humans, particularly during the hot summer months. This is a wise bit of folk wisdom but fortunately, with the coming of modern refrigeration methods, the old prohibition no longer is valid and oysters can be eaten throughout the year.

Q *Does acupuncture really work?*

A Many centuries ago, the Chinese noted that warriors struck by an arrow, say in the shoulder, were cured of arthritis in that shoulder. From such beginnings, acupuncture developed to the point (no pun intended) where a thousand places on the body can be penetrated by a needle to cure just about everything. This practice spread to the West, especially France and Germany, where many cures were said to have taken place. Undoubtedly, cures have occurred, but either the ailment would have gone away anyway or the patient had unbounded faith in the value of acupuncture and so influenced the result. There is no physiological basis we know of to support this method of treatment. It is very difficult to accept that a needle in the right toe will cure a headache. Actually, the Chinese believe that acupuncture rests on the supernatural basis of the life forces yin and yang acting within the body.

Q *How common is the belief that it is unhealthy to touch water or something cold when the body is overheated?*

A We can't say how common it is but the theory of hot/cold distinctions does exist and takes various forms. In many societies, it is considered dangerous to drink or make any contact with water if the body is overwarm—for example, overwarm from using a hot instrument such as an iron, from walking across a hot dry field, and from similar activities.

The hot/cold distinctions, by traditional custom, apply also to foods as they may affect health. In this theory of folk medicine, foods are classified as having "hot" or "cold" properties. Bread, sugar, garlic, and brandy are "hot" while beans, milk, and oatmeal are examples of "cold" foods. The wrong mix, according to folk belief, can cause temporary or acute illness.

The hot versus cold imbalance theory is much older than modern medical science, being traceable back at least as far as Greek and Roman times.

Q *Why is Popeye supposed to get strong after eating spinach?*

A There are many food myths, of which this is one. Most myths are rooted in some factual soil more fertile than is warranted. Around the turn of the century, a widely circulated nutrition book contained a slight typographical error. A decimal was misplaced in the value of iron in spinach so that it supposedly contained ten times more iron than it actually did. Iron is associated with strength and endurance. This may have been the basis for Popeye's legendary strength.

Also of interest is the reason why carrots are thought to improve eyesight. During World War II, the British had developed radar, unbeknownst to the Germans, and were knocking out enemy ships and planes with uncanny accuracy. To keep the Germans off the track of developing radar, the British circulated the story that their pilots were eating giant carrots that miraculously improved eyesight. Whether the Germans fell for this story is not known, but the idea that carrots are good for the eyes has persisted ever since.

Q *How did people keep food from spoiling before the era of modern refrigeration?*

A A lot of food was not kept from spoiling. And especially with primitive tribes, the only solution was to have feasts where everybody gorged themselves before the food spoiled. However, it wasn't too long before salt was found to be a great preservative, particularly for meats. Other foods that could not be preserved by salt were kept in cold-water streams. On farmsteads of the last century, streams were sometimes diverted through the basement of the house and food kept in the cool water. This was especially effective if the streams drained snow-covered mountains. Another way to keep meat from spoiling was to keep it alive and "on the hoof" until it was ready to be eaten. Notice that our concern for food spoiling, even with refrigeration, extends to the present. Markets continue to tell the consumer how fresh the food is in many ways: fresh-caught fish, strictly fresh eggs, home-grown produce (meaning of local origin and thus still fresh), live lobster. It may well be that food spoilage in the remote past signified disaster in terms of survival and that that feeling is still with us.

Q *Will science ever find a way to eliminate the need for sleep?*

A We have serious doubts, based upon the evidence. Certain chemicals can be taken to prolong the awake period but sooner or later, the individual will crash and in fact it is not healthy to do this. In experiments designed to keep persons awake for long periods, nervousness, hallucinations, and plain psychotic behavior resulted. In other words, a person deprived of sleep behaves like a crazy person. As you probably know, sleep-deprivation tactics have often been used against prisoners of war in an attempt to make them talk. When people are allowed to sleep after long durations without sleep, they experience more frequent REM (rapid eye movement) periods of sleep. These are the periods when dreaming occurs. It would appear that dreaming and sleeping are critical for recharging our batteries, so to speak. If science can get around this obstacle to sleep-free living, it is not yet on the horizon. Anyway, what's wrong with a nice snooze?

Q *I keep dusting in my house and it keeps coming back. Where does all this dust come from?*

A By definition, dust consists of particles less than a millimeter in diameter, whatever the composition of these particles, and they may arise from a number of different sources. The dust in an average house might have a significant amount of dead skin fragments because your skin is constantly shedding little flakes. Tobacco smoke is another source. Open windows and doors admit windborne particles from the atmosphere, including pollen or other plant particles, industrial smoke particles, and clay or other mineral matter from soils.

Children parading in and out are a rich source of dust. Major volcanic eruptions produce thousands of tons of dust, which eventually must settle somewhere, and even quantities of dust from outer space should not be overlooked. About 43 million tons of dust settle over the United States each year. Of this amount, 31 million tons is from natural sources and the rest is from human activity. Like it or not, we will have to learn to live with dust.

Q *It is true that noise can be a health hazard?*

A Noise is definitely a health hazard and can be regarded as a form of pollution. Fifty percent of the U.S. population is exposed to a noise level that interferes with speech or sleeping. Sources of noise include cars, motorcyles, lawn mowers, jet aircraft, and even barking dogs. This can lead to severe mental disturbances, even provoking violence, as when a man in Miami, Florida, shot his neighbor because of a loud stereo. Noise is measured in a unit known as the decibel. Normal conversation is 60 decibels. A loud stereo puts out 120 decibels, which can lead to impaired hearing. At 140 decibels, actual physical pain may be experienced. Oddly enough, loud noise is considered ‘‘masculine,’’ one example being a loud motorcycle. Without moralizing, we suggest that everyone ‘‘keep it down.’’

Q *Since the 1950s, a lot has been heard about the ecology movement, particularly how man has disturbed and polluted the natural environment. How long has man done this?*

A Man, of necessity, has always altered the natural environment to some extent. In the earliest prehistoric societies, the effect on the environment was slight because there were few people and they needed from nature only food, drink, and shelter.

Later, man began to change and control the natural environment, sometimes leading to ecological disaster. As long ago as 350 B.C., the Greek philosopher Plato noticed that one effect of Greek civilization was the extensive cutting of forests to meet the demands for fuel, grazing land, shipbuilding, and other construction. As today, deforestation led to erosion and other negative effects on the ecosystem in Greece over 2000 years ago.

The urbanism of the Roman empire, open pit mining, and faulty agricultural practices led to air, noise, and water pollution in the ancient city of Rome and even then laws had to be passed to guard public health. In other words, bumper sticker ecology is nothing new.

Q *What is the best way to dispose of hazardous nuclear wastes?*

A We wish there was an easy answer but there is not. Some scientists favor subsurface disposal in salt beds, but any incursion of groundwater may dissolve the salt and release the wastes into our groundwater. Others believe they should be stored at the surface in safe containers where they can be more carefully watched, and in addition these wastes would be available if a use were found for them, which is possible. One suggestion that seems to make sense is to store them at nuclear test sites where the land has already been contaminated and can't be used anyway. To dump them into the deep part of the oceans is not satisfactory because leakage would contaminate the sea and our goose would really be cooked. A major problem is that nuclear wastes can remain dangerous for thousands of years, and whatever final action we take now could affect many future generations.

Q *Because the oceans are so vast, is there really a serious threat that they may become polluted?*

A While many substances reaching the oceans will be greatly diluted and rendered harmless, the oceans are still vulnerable. Keep in mind that the oxygen in the atmosphere that we breathe and otherwise use up is replenished mainly by the phytoplankton in the oceans, not by land vegetation. Even tiny amounts of, for example, DDT have adverse effects on oxygen production by phytoplankton. Another problem is oil spills. Tankers and other vessels add 3.5 million tons of oil to the oceans each year, which affects marine life. One possible danger is the rupture of underwater oil pipelines. There are 20,000 miles of these pipes, and they are slowly rusting. They will have to be watched closely.

4
MAN'S PAST

From Old World to New World

Q *Is man the only creature that makes war and kills members of his own species?*

A When its own life is threatened, or that of its offspring, an animal, regardless of species, as far as we know, will attack and kill its own species. Man is certainly no exception to this. Among the mammals, a male may attempt to kill the young because they are seen as a future threat to the adult male's superiority. The mother will defend the offspring to the death. This is fairly obvious. With man, it may be less obvious when two countries go to war and soldiers die. Yet on a more sophisticated level, the war is fought because a threat to survival is perceived.

Ants kill other ants and even take slaves.

Q *According to scientists, how long has man been on the earth?*

A It is amusing that scientists argue among themselves about what should be considered human. Skeletal remains found in the Middle East are certainly human and date back 60,000 years. Beyond that, scientists are coming around more and more to the idea that Neanderthal man should be considered one of our species. They were intelligent, had complex rituals, and buried their dead. They have now been given the name *Homo sapiens neandertalensis,* and this new classification of Neanderthals gives the human lineage an age of at least 150,000 years.

Q *I see references to the Stone Age, the Paleolithic, the Bronze Age, and so on. Can you sort these out for me?*

A Once a man appeared on the earth, he began to make tools, such as for hunting. Later, he became more knowledgeable and skillful. Scientists have been able to establish a chronology of early man's development. This development has been categorized as follows. The Stone Age was a period of making and using stone implements. It can be subdivided into the old Stone Age (the Paleolithic) and the new Stone Age (the Neolithic). The Paleolithic extended up until about 30,000 years ago and involved the making of crude stone tools. From 30,000 to 10,000 B.C. was the Neolithic, during which man made more sophisticated stone tools; this was also the time of cave paintings and elaborate burials. Subsequently, man started to use metals rather than stone, ushering in the Bronze Age about 12,000 years ago, and after that, the Iron Age.

Q *Why has man survived this long, considering the number of wars he has begun?*

A It may actually be that our aggressive traits were instrumental in our survival. Consider other animals.

Many species of birds, reptiles, mammals, and amphibians show aggressive traits. This is seen in the animal's staking out an area or territory containing the resources needed for survival. This territorial instinct intensifies during both mating season and the rearing of the young. Some animals stake out large areas and offer combat to those that invade, sometimes to the death. Some, such as a bird, may only consider the nest itself as the defended territory. Where animals are not aggressive, we may see extinction. An example is the passenger pigeon, which offered no resistance to egg hunters destroying their nests, and thus was wiped out in this century. Aggression in man can be seen as normal. Our problem is we keep dreaming up more and more efficient ways of expressing this aggression.

Q *Archaeologists often have to dig below the ground to find ruined buildings. How did they get covered? Has the ground level risen?*

A Archaeologists have to excavate because the workings of nature help to conceal ruins, fossils, and other relics. For example, consider an abandoned barn. Collapsed from neglect, the wood begins to decay. Wind and rain bring in soil whereupon weeds and trees take root. The plants discard leaves and die. This organic debris piles up, a new soil may form, and over time, the barn becomes completely buried. Other buildings may be burned, torn down, or partly destroyed (in war, for example) and new structures built on top of the old. The old land surface then becomes artificially buried. A case in point are buildings of the Aztec civilization found below the streets of Mexico City.

Q *Why has Neanderthal man been depicted as being able only to grunt and snarl?*

A Poor Neanderthal man has been much maligned. The popular concept of a shuffling, bent creature is due to faulty analysis of skeletal material. Neanderthal man was not that different physically from modern man, but different enough to not be mistaken for him. Recent studies of the oral and throat skeletal cavities suggest that Neanderthal man was not well equipped to speak as we do. This may well have been a contributing factor in his disappearance 30,000 years ago. Hand signals plus inefficient oral signals would not be as useful as full control of a spoken language, even one of limited vocabulary. Thus, Cro-Magnon man gained superiority and indeed may have been the one to have wiped out the Neanderthals. A few scientists have suggested that Neanderthals interbred with the ancestors of modern man. The Neanderthals were hardy, aggressive, and survival-oriented. Perhaps we carry Neanderthal traits and for that reason have been successful at surviving.

Q *Was Neanderthal man a vicious, predatory creature that tried to wipe out real humans?*

A It is ironic that the question should be so phrased because if anything, it was probably the other way around. We know that Neanderthals lived in northern Europe between 100,000 and 30,000 years ago. Their brain capacity was actually greater than that of modern humans and there is no evidence to suggest that they were any more vicious than we are. Anatomically modern humans (the Cro-Magnons) invaded western Europe about 40,000 years ago and came face-to-face with the

Neanderthals. In 10,000 years, the Neanderthals were gone.

It is uncertain whether Neanderthals were killed off, interbred with the Cro-Magnons, or a combination of both. The midfacial prognathism of some western Europeans gives some support to interbreeding. If you are a descendant of a west European, it is likely that the blood of both Cro-Magnon man and Neanderthal man runs in your veins.

Q *How long have people believed in ghosts, and for what reason?*

A An exact date is not known, but belief in ghosts must go back several thousand years, even to the most primitive cultures. The reason we say this is because ancient burials are known in which the corpses were tied down or buried head first, and with heavy rocks over the grave, suggesting the belief that a corpse might rise and walk again. We can speculate that comatose or otherwise unconscious persons were thought to be dead, and then seemed to come back to life. It would have been particularly shocking if an unconscious person were buried in a shallow grave, revived, and managed to crawl out. Early man saw death and didn't like it. Belief in ghosts may well have provided a sign of immortality, as it does to many people today.

Q *How old is the earliest known writing?*

A The earliest writing we know of is called cuneiform and is at least 6000 years old. These were markings made on clay tablets by the Sumerians, a people who lived in Mesopotamia. Others such as the Assyrians also used this method of writing, which took the form of wedge-like markings with various orientations. In some cases, we must admit we don't know how to read it. Historians and archaeologists believe that writing was invented to keep track of tribute and other deposits of money that farmers made to the city-states of Mesopotamia, from which grain and other produce were distributed. It is probably safe to say that the first writers were priests because these early kingdoms were very religion-oriented.

Q *Are there, or were there, subterranean races that lived in deep caves thousands of years ago?*

A Early man used caves as shelter and refuge from the elements and savage beasts, but these caves were mostly at or near the earth's surface. Legends of "mysterious" races in caves may have started with the discovery of Mammoth Cave in Kentucky in the late 18th century. Mummified remains of humans, thought at that time to be pre-Indian, were found, and these crumbled to dust when touched. However, reed torches have been found, and thus the bodies are probably those of Indians who, exploring the cave, got lost and died, not surprising when you consider that Mammoth Cave has more than 150 miles of passageways and galleries. The idea posed in the question is quite preposterous. Nonetheless, caves are intriguing, and any newly discovered one may offer surprises.

Q *Flies and mosquitoes drive me crazy, but we have sprays. What about primitive man? Did they have some special way to cope with this problem?*

A Anthropologists have no particular answer to this except to say that early man enjoyed the attention of these insects no more than we do today. Even now, in less advanced countries smoke pots are still used to drive the insects away. Consider also that we humans seem to be able to get used to anything we regard as a part of our lives. We have seen people in less advanced societies covered with various kinds of bugs around the head and body, yet paying little attention. Yet those of us who do not usually experience this kind of attention and are not used to it can spend an entire night chasing one mosquito around the bedroom. Let's reverse the coin and think of an average New Yorker dodging early morning traffic on his way to work. Ho hum, he thinks, it is a part of life. But put a caveman on Times Square and he'd go insane. It is a case of adaptability, whether pleasant or unpleasant, and our species has proven very good at that.

Q *When did man begin to farm?*

A No one can be certain. Domestication of plants seems to have taken place in the Old World in the Near East and in the New World in Central America, perhaps Mexico. So it is a case of what archaeologists call independent invention. Popular wisdom used to be that agriculture began about 10,000 years ago. However, more recent findings now lead us to believe that the raising of crops could have occurred 18,000 years ago in the Near East. Speculation as to how this practice began is as follows. The earliest humans were hunters and gatherers, always on the move. Perhaps one lush area of abundant food supply allowed a particular group to stay long enough to observe that casually discarded seeds from gathered fruits or vegetables germinated and produced more of the same. Wheat and corn were the first crops. The goat was the first animal domesticated as a farm animal.

Q *Which did man domesticate first, dogs or cats?*

A Skeletons of dogs have been found in the remains of human habitations dating back at least 10,000 years. Dog experts believe our modern dogs are derived from the gray wolf, which roamed all over Europe in primitive times. We can imagine as the starting point a caveman finding a litter of orphaned puppies and taking them home. Cats may have been domesticated somewhat later because dogs, as hunters, would have been more useful to early man, also a hunter. We do know that cats had become pets of the ancient Egyptians, who mummified them and even had cat cemeteries 5000 years ago. Indeed, the Egyptians worshipped the cat as a sacred animal. The present-day diversity of dogs and cats is due to man's own breeding experiments, one of the prime lines of evidence for organic evolution.

Q *I was eating a slice of bread and wondered who first thought of it—and I'm not even sure how you make bread.*

A Consider that more than 20 billion pounds of bread are consumed in the United States alone each year—quite a number of the consumers probably don't know either. Archaeological evidence shows that bread was made at least 10,000 years ago in the region that is today

Switzerland. This was not our modern bread, but rather ground acorn mixed with water and heated to form cakes. Later, bread was made with ground cereals such as wheat, rye, barley, and corn. The ancient Egyptians were apparently the first to make white bread from wheat and "raise" it by using yeast. Archaeologists have found ancient ovens where the bread was baked. With regard to bread's "invention," accident likely played a major role, as it did in fire and cooking, the wheel, and the bow and arrow. The bread that is eaten today was a luxury reserved for royalty as late as the 17th century; commoners ate the flat, rather tasteless bread similar to that made perhaps thousands of years before.

Q *Do the Stone Age aborigines in Australia believe in life after death?*

A They have a very elaborate system of beliefs, including reincarnation and a form of Supreme Being in the sky. We do not know what religious beliefs they held when they first arrived in Australia 20,000 to 30,000 years ago, perhaps from the area of Southeast Asia, but the environment of Australia came to shape their beliefs. Water is scarce in that country and thus a watering place is essential for survival. The aborigines believe each water hole is a dwelling place for spirits of their ancestors, waiting to be reincarnated. Their religion also identified closely with animals and even inanimate objects such as rocks and boulders. Their rituals include painful rites to show manhood, such as scarring of the body and occasionally knocking out a tooth with a hammer.

Q *Did ancient peoples believe in one supreme God?*

A No. Early peoples believed in many gods, each of which had control over a separate aspect of the environment. The idea of a single all-powerful god is actually the most modern of all religious ideas. The belief in many gods—polytheism—seems to have taken root in ancestor worship, where departed loved ones became protectors watching over a family or tribe from a spiritual plane. Other gods personified different parts of the natural environment—such as wind gods, river gods, earthquake gods, and so on.

The first person to express an official position for a single all-powerful god seems to have been the Egyptian Pharoah Akhenaton, who perceived the sun as the personification of that god. There is suspicion that he was later murdered for this belief. Certainly not too long afterward the Israelites proclaimed one God for the Jews, but the idea that there was one universal God for all people seems to have become entrenched during the time of the early Greeks.

Q *Are the Chinese the only ones who practice ancestor worship?*

A By no means. It is still an important aspect of many different cultures around the world, especially in modern primitive societies in Africa, parts of the Pacific, and India and Japan. We see evidence as far back as Neanderthal man, perhaps earlier, that there was belief in survival after death. Ancient graves often were well equipped with food, weapons, and other possessions to aid the dead in their second life. In present-day societies, at regular intervals food is placed at the shrines of ancestors. The ancestors, although usually benevolent toward the descendants, may at times punish them or become angry for some reason. Thus, besides affection and respect for the dead ancestor, there is often fear. In a few societies, human sacrifices have been made to the revered ancestor. Even in a modern society such as ours, the deceased are revered and idealized in a way that approaches worship. In a sense, Christian saints represent a form of ancestor worship.

Q *Is it true that the cross, far from being a Christian symbol, was a pagan device long before Christ?*

A Crosses have been found painted on pebbles, at archaeological sites more than 12,000 years old, and thus would have had nothing to do with the Crucifixion. It is believed by scientists that the cross is one of those universal symbols signifying the four main directions of the compass, and so represents all of life and everything. We know that the early Egyptians used a similar symbol with a loop at the top to symbolize life. It is thus entirely consistent that the cross also be a symbol of Christianity.

The question posed above implies that a pagan origin is degrading. By definition, a pagan is a person who does not share the beliefs of Christians, Jews, or Moslems. Early on, Romans and Greeks were considered pagans because they believed in more than one god. While we might argue today at that illogic, the Greeks and Romans, as pagans, contributed significantly to science, art, and the advancement of civilization as we know it today.

Q *When did man first start using pottery and what significance does it have?*

A To an archaeologist, pottery can sometimes be a discovery more important than that of a chest of gold coins. For one thing, pottery has been around a long time, some even being about 10,000 years old. This means that it is also very durable. Pots break, of course, but the pieces—called shards by archaeologists—do not disintegrate even over thousands of years. Before firing (the process that makes pottery almost like stone), the clay can be shaped and decorated in a virtually unlimited number of forms and styles.

A very important consideration is that pottery styles changed through time and this provides a means of estimating age. Styles are datable in the same way that automobiles can be dated according to their design.

Although we usually think of pots, fired clay was also used for jewelry, lamps, and construction material, and even as tablets for the earliest form of writing. Thus, pottery is for the archaeologist one of the single most important remains from the past.

Q *Today, archaeologists use ancient pottery as the main evidence of early peoples. Thousands of years from now, what do you think they will look for from our civilization?*

A Our civilization is far more widespread and complex than earlier ones and so the number and type of artifacts would be greater. Also, opportunities for preservation would be greater because a not insignificant part of our civilization is underground and protected from destructive elements. We may assume that a few bank vaults, the below-ground levels of large buildings, and subway systems would be preserved. With time, objects made of iron would crumble away, but many of the things left over from earlier civilizations would likewise persist, such as glass, other ceramics, and bricks, as well as concrete, some plastic, and wood. Tapes and film might not last, but the written word is so extensive that we imagine future archaeologists would soon translate our languages. Another factor involved is that these archaeologists may well have very sophisticated excavation and preserving techniques that will reveal more than present methods. If it is an atomic catastrophe that destroys us, much less would probably remain.

Q *Can it be that ancient batteries found near Baghdad prove that the advanced technology of aliens was available to primitive peoples?*

A It is true that primitive batteries capable of generating one-half volt of electricity have been found, not at just one, but at more than a dozen archaeological sites. These would be useful in electroplating silver onto copper. Such devices are between 2200 and 1800 years old. We are faced with the following choices: (a) people back then were intelligent enough to construct such a battery, or (b) an advanced civilization reached the earth in spaceships and showed us how to make a ridiculously crude battery. We like to think that humans 2000 or even 5000 years ago were as smart as we are and could make a battery. After all, it is not a sophisticated battery and would hardly be a type of battery aliens would bring with them. When one delves into ancient technology, it is remarkable how skilled early peoples were in mechanics, art, transport, and science.

Q *Is the Great Pyramid in Egypt the largest man-made structure in the world?*

A Despite its magnificence, with more than two million limestone blocks used in its construction, it is not. That prize would have to go to the Great Wall of China. It is 1700 miles long, and would stretch from New York City to Topeka, Kansas. There is enough stone in the Great Wall to build 30 Great Pyramids. Construction began in the second century A.D. and continued over the next several centuries. The wall was intended to keep out invaders from the north, but failed in this purpose when it was breached by the founders of the Manchu dynasty. Remarkable is the fact that the wall was built up and over steep mountains, so that in places it is like climbing a step-ladder to walk the wall. It is considered the longest graveyard in the world because slaves used for its construction were buried in or near the wall when they died.

Q *Were mysterious ceremonies conducted inside the Great Pyramid in Egypt thousands of years ago? There are air vents leading to the King's Chamber, and the dead have no need of air.*

A There is some speculation that this may have been so, but archaeologists maintain that the primary function of the pyramids was as burial sites, particularly of the royal houses of Egypt. However, this does not preclude the use of these great structures for other purposes. For example, during the construction of the Great Pyramid, it may have served as an observational platform for the astronomers. Intriguingly, there is a shadow floor on the north side of the Great Pyramid. By marking daily the position of the shadow cast by the Great Pyramid, a calendar of sorts could have been constructed. In that sense, the Great Pyramid might have been a giant sundial. Purposes such as these are obscured by time, and we may never have clear answers to such mysteries.

Q *Have scientists employed instruments in an attempt to find hidden chambers in the pyramids?*

A Yes, but in those man-made mountains, the Egyptian pyramids, it's not easy even with modern sensing devices. Some archaeologists have suspected a second hidden chamber in the pyramid of Chephren at Giza.

Chephren is thought to have been the son of Cheops and the two largest pyramids in the great complex near Cairo are those of father and son.

American physicists developed an instrument to send radio emissions or pulses through rock and to record any changes in density, such as would be caused by the hollow space of a tomb. No chamber was found, but the remote sensing technique may someday aid archaeologists at another site.

Q *Does the Great Pyramid contain mysterious mathematical truths?*

A Built more than 4000 years ago by Cheops, a fourth dynasty king of Egypt, this enormous funerary building has been used as the basis for many popular legends and claims, including its power to predict the future, preserve food, cure disease, and so on. One theory is that measurements of the Great Pyramid reveal that ancient Egyptians had a profound mathematical knowledge of the universe.

It has been said, for example, that the builders used a special unit of measure, the "pyramid inch," and that the height of the pyramid in inches, multiplied by one billion, would equal the distance between the earth and the sun. And by calculating the volume of the pyramid in cubic pyramid inches, the result would equal the total number of all people that had lived on the earth since creation.

Although such notions do stimulate our imagination, archaeologists have found that these arguments are often based on inaccurate measurements, and there is no evidence of any cosmic significance in the Great Pyramid. On the other hand, Cheops's pyramid is recognized as an amazing engineering and organizational accomplishment.

Q *Was there really a death curse on all who desecrated the tombs of the Egyptian pharaohs?*

A This idea got rolling when King Tutankhamen's tomb was discovered on November 3, 1922, by archaeologist Howard Carter. Lord Carnarvon, who bankrolled the expedition, died within two months of the opening of the tomb. This was followed by the deaths of the two men who photographed and X-rayed the mummy. The British archaeologist H. E. Evelyn-White committed suicide, leaving the cryptic note, "I knew there was a curse on me," adding strength to the legend.

If Tut was out for revenge, he did a lousy job. The chief despoiler, Howard Carter, lived to age 67, and died in 1939. Other archaeologists such as Flinders Petrie and Edward Newberry, both of whom worked in the tomb, even eating and sleeping there, lived to ages 89 and 80, respectively. While the deaths of some members of the expedition remain a mystery, the belief in a curse is without solid proof.

Q *Where did the Egyptians get the idea for mummifying their dead?*

A The Egyptians, who were probably the earliest to do so, carried the art of embalming bodies to a high degree for several thousand years during the Dynastic period. The practice was based on the belief that the deceased's identity would endure if his body and features were preserved. Although the ancient embalmers learned how to effectively use resins and salt-containing crystals, the heat and dryness of the north African climate itself acted as a good preservative.

Much earlier, in the Predynastic period, bodies buried in shallow desert graves became mummified by the natural desiccation and this probably influenced the practice of mummification in Egypt.

Q *Why did the Egyptians mummify animals?*

A Since we know that the Egyptians believed in a kingdom of the dead and wanted to prepare themselves for the life after death by mummification, we might assume that they simply wanted their pets along.

The fact is though that animals were mummified because they were considered sacred. The god Horus, for example, is portrayed with the head of a falcon, and other animals such as the cat and jackal were likewise sacred to other gods.

Dogs, bulls, owls, fish, and even reptiles were mummified and given royal burials, sometimes in coffins shaped to fit them.

Q *The word "mummy" is used to describe a wrapped and preserved body. How did this word originate?*

A It seems to have been derived from a Persian word, *mumis,* meaning wax of bitumen. Centuries ago, it was believed that waxes, bitumen, and some oils had medicinal value and these were applied in an attempt to cure some ailments. In ancient Egypt, the people believed that immortality could only be attained if the body was preserved. In the preparation of the deceased for burial, bitumen was applied liberally, and even packed into bodily cavities after certain organs were removed, the thinking being, we suppose, that if it was good enough to preserve the living, it was good enough to preserve the dead. The word "mummy" thus came to indicate a corpse preserved in this fashion.

Q *What do Egyptian mummies look like unwrapped?*

A Quite often the features of the person are well preserved and include the nose and ears. The Egyptians believed that preservation of the body was essential to ensure immortality in a second life. Although they did attempt to develop preservation techniques, most of their success was due to the dry climate in Egypt, which would tend to preserve tissue in any event. Indeed, some of their techniques were counterproductive, as in the case of using asphaltic oil as a preservative. They applied this to the corpse of King Tut and he looks simply awful. Unwrapping of a mummy is a painstaking business because the oil they used became impregnated into the wrappings and over the centuries was lithified. Yet study of these mummies provides useful information about the health history of the individual and the cause of death. Some mummies can be studied without unwrapping by using X rays.

Q *Did other people preserve their dead as mummies?*

A Yes. There were mummies in many countries but perhaps not as old as the Egyptian mummies, which date back 6000 years. In the Canary Islands, the dead were preserved in almost exactly the way used by the Egyptians of the 21st dynasty. This included a flank incision to remove internal organs for separate preservation in jars, and tightly bound bandages around the body. The Incas used similar techniques; their mummies can be found in Ecuador, Peru, Venezuela, and Bolivia. Sometimes they painted the mummies with red ocher. Mummification practices also existed in Australia. From the Middle Ages to the 18th century it was widely believed that mummies had great value as a medicine and were imported into Europe for sale. Because mummies were rather rare, fake mummies, actually the bodies of executed criminals, were doctored to look like mummies and sold. Basically, in all cultures that practiced mummification, it was believed that the person would enjoy a second life if the original body was preserved.

Q *No mummies were found in the pyramids in Egypt so they could not have been tombs. Doesn't this contradict the claim of the archaeologists?*

A We would have to say no. Maybe you have been swayed by those who insist that the pyramids were built to focus cosmic force or to house flying saucers or other fabulous claims. Bear in mind that, while there are only seven outstanding pyramids in Egypt, there are dozens of other less notable pyramids and these did indeed shelter mummified remains of Egyptians. The famous pyramids, such as the Great Pyramid of Khufu near Cairo, didn't, it is true, contain the corpses of pharaohs. Why? Egyptians were deeply religious and believed that the body must be preserved, along with its treasures, in order to enjoy a second life. But the pharaohs were not unaware of grave robbing, which according to their beliefs would deny them this opportunity. Archaeologists suspect that some pharaohs may have built pyramids as a sort of "decoy," while being secretly buried at an inconspicuous site in the desert. Nonetheless, some grave robbers did indeed gain entrance into the tombs and took anything of value, including the mummies.

Q *What exactly does the Egyptian sphinx represent?*

A The sphinx represents the Egyptian Pharaoh Chephren, who ruled during the Fourth Dynasty, about 4500 years ago. It has the body of a lion and a human head, probably showing the strength and power of the ruler. It was defaced by Napoleon's soldiers, who blew off the nose with a cannon during target practice. A sphinx can be seen in the art forms of many peoples, particularly in the Middle East. The Greek sphinx was a woman with the body of a lion. Sometimes the sphinx also has wings.

Q *Where was Cleopatra buried?*

A There were various women in ancient history named Cleopatra but you probably have in mind the most famous of these—Cleopatra VII. Born in 69 B.C., she ruled as a queen in Egypt during the period of the Ptolemaic or Macedonian Greek occupation of that country.

Although probably not as beautiful as her film version, played by Elizabeth Taylor, Cleopatra was charming, intelligent, and courageous and she came to have a strong influence on the lives of Roman statesmen, among them Julius Caesar and Mark Anthony. Under the Ptolemaic dynasty, the coastal city of Alexandria became a great international center and eventually the target of Roman expansion. It was there that Cleopatra, last of the Ptolemies, died in 30 B.C., after which Egypt became a Roman province. History records that she had constructed for herself a mausoleum in the royal cemetery at Alexandria. According to popular belief, she chose to take her own life by snake bite.

Q *Why did Cleopatra commit suicide by snake bite?*

A Cleopatra, Queen of Egypt, was the wife of Mark Anthony. In the civil wars that wracked Rome at that time, Octavian had managed to defeat Anthony, who thereupon received a false report that Cleopatra had committed suicide; hearing this, he took his own life. When Cleopatra learned of this, she too committed suicide. She chose to die by snake bite because the snake was sacred in Egypt's religion. It is of interest that Cleopatra was not Egyptian, but Greek, with some Iranian blood. Portraits of her exist that show she was not as beautiful as some say, but had such charm, personality, and intelligence as to be regarded as a great historical personage.

Q *What is known concerning the origin of the practice of wearing earrings?*

A Archaeologists have been digging up ancient earrings for a long time and conclude that they have been used for many centuries in many different cultures around the world. Greek statues from the fourth century B.C. show pierced earlobes, so this practice was probably the most functional way to wear earrings, which consisted of gold or precious gems. The use of clamping devices to hold earrings in place is a 20th century custom and not as popular as piercing the ears. We can see two reasons why earrings have been worn: as decoration and for protection. The ears are second only to the eyes in sensory importance. Early superstitious beliefs included the idea that an evil spirit could enter the body through the ear, and so earrings were worn as an amulet or talisman to protect against this. Later, sailors wore gold earrings as protection against drowning. In some societies, the lobes were enlarged into loops to carry practical objects such as sewing kits, and this practice continues. For the most part, it was women who wore earrings, even many centuries ago.

Q *How did the idea of biblical creation start and become so widespread?*

A In the Christian tradition, it is believed by most scholars that Moses wrote the first five books of the Old Testament, which perhaps were later modified in the fifth century B.C. by what was called the Priestly School. These books of course include Genesis. Many non-Christian cultures—such as the Babylonians, Egyptians, and Indians—have their own creation stories. These mostly predate the Christian story of creation by as much as 1000 years. Many scholars have noted the strong influence earlier accounts had on the Christian version. It is not surprising that early peoples made some attempt to explain how the earth was made and how humans came on the scene. Such concern over creation seems to have extended back into Paleolithic times. Following the time of Christ, Christianity spread rapidly into many countries and then to the New World. Naturally, the Bible story accompanied the spread of Christianity.

Q *How could Methuselah and others in the Bible have lived so long?*

A Chapter 5 in Genesis mentions Methuselah's age as 969 years while Adam lived to be 930. Many other persons in the Bible had comparable longevities. It is difficult for science to reconcile such claims with what is known about human longevity today or even in the human fossil record extending back many thousands of years. One answer might be that during biblical times, time was measured differently. For example, if a month were equated to a year, then Adam's life span would be 77 years. This is reasonable. The idea that the earth traveled faster around the sun in those days and made for a shorter year is unsupported by science. This is one of those secrets or hidden meanings of the Bible for which we may never be able to find an answer.

Q *Is there any natural explanation for the parting of the Red Sea during the Exodus of Moses and the Israelites?*

A Several explanations have been offered other than the direct and miraculous intervention of God. Some historians claim that the refugees actually crossed at a point of soft marshes and not open water and the idea that walls of water were formed is just fantasy. Then, in the 1950s, Velikovsky proposed that a near collision with the planet Venus triggered tidal eruptions that drained away the Red Sea temporarily until the Israelites could cross. The waters then swept back when the pharaoh and his army attempted the crossing, and they were drowned as the Bible claims. There is no scientific foundation for the theories of Velikovsky. However, at that time 3500 years ago, one of the most violent volcanic eruptions of all time took place—the eruption of Thera. It must have darkened the entire Mediterranean and sent giant sea waves (tsunami) through the area, and this might have achieved the same results: the temporary parting of the Red Sea. It would also account for the various plagues of Egypt as recounted in the Bible.

Q *A movie has portrayed the search of the Ark of the Covenant. Was the movie accurate?*

A Commercial films offer exciting but often unrealistic versions of archaeological excavation. In this case the object of the hunt, the ark, refers to a chest of wood covered in gold and containing tablets of the Ten Commandments, which, according to the book of Deuteronomy, marked the covenant between God and Israel. By tradition, the ark was the most sacred religious symbol of the Israelites. It was sometimes carried in battle or kept in sanctuaries such as the innermost chamber of King Solomon's temple in Jerusalem. We do not know when or why this original ark disappeared. Ancient synagogues contemporary with the Roman and Byzantine empires had versions of this ark but these were used to store the holy scrolls of the Jewish people including the Law or Torah when not in use, not the tablets of the Ten Commandments. The niches in which these "arks" resided are often found in the excavations of the ancient synagogue buildings, but not the boxes themselves. There are no death curses, snakes, or other hazards created by Hollywood associated with these excavations.

Q *Is it true that the Jews committed suicide rather than surrender to the Romans?*

A Yes. Masada was a strong fortress built on a steep hill in about 100 B.C. although it may have been used by wandering groups before that time. It is located along the coast of the Dead Sea in Israel. It was refurbished by Herod in 35 B.C. and remained a Roman fort until 66 A.D., when it was seized by Jewish zealots who were revolting against the Romans. The revolt lasted until 72 A.D. when Masada was the last stronghold still in the hands of the Jews. The Roman tenth legion laid siege there and the 960 defenders took their own lives rather than surrender. Archaeologists have found important scrolls there in the course of their excavations.

Q *Does any part of the Cross of Christ's Crucifixion actually exist?*

A This is a tough one. According to legend, Helena, the mother of the Emperor Constantine, went to Jerusalem in the fourth century and directed excavations at the site of the Crucifixion. She found three crosses buried and identified the Cross by means of its healing power. She supposedly also found the nails. What is troubling about this story is that the Cross was buried. In those days, crucifixion was a common form of capital punishment, and usually meted out to lower classes such as slaves. We do not know why the Romans would have buried the crosses they used for execution. It does appear, if it was the Cross that was found, that it was sliced into slivers and distributed to churches throughout Europe. Such a fragment is said to be housed in the basilica of Santa Croce in Rome. It is unfortunate but true that if all the wood around the world said to have come from the Cross were gathered together, there would be enough wood to build a ten-room house.

Q *What is known concerning the Shroud of Turin?*

A The Shroud is the purported original linen in which Christ's body was wrapped. The image of face and body impressed upon the cloth is supposed to be the only "photograph" of Jesus. The weave does date from the first century and pollen types found in the cloth are from Palestine. The image seems to have been produced by a chemical reaction between ointments (myrrh and aloes) and ammonia given off by the body. It is without a doubt the portrait of a crucified man about five feet ten inches tall with a beard, a crown of thorns, wounds in wrists, feet, and sides, and scourge marks on the back. The recent opinion of some who believe in the authenticity of the Shroud is that Christ, contrary to popular belief, was blond-haired and blue-eyed.

Skeptics still point out that thousands were crucified in

Christ's day, so there is nothing unusual about the image being that of a crucified man. Also, the Shroud was little heard of until the 14th century—a time when the art of "rubbing," a process that could produce such an image on cloth, was well developed. We will still have to wait until radiocarbon dating of the cloth establishes its true age.

Q *How is it that the Romans persecuted the early Christians and now Rome is the center of Catholicism?*

A In 312 A.D., Constantine was struggling to gain control of the Roman Empire. The Christians were a small and unimportant minority suffering periodic persecution. Constantine had seized Gaul and invaded Italy to attack his chief rival, Maxentius, who had a large army. On the eve of the battle, Constantine had his soldiers paint a cross on their shields. The forces of Maxentius were defeated and Maxentius was killed. Thus, Constantine became master of the Empire and proclaimed Christianity as the official religion. Years later, Constantine told his biographer that before that battle, he had seen a cross of light in the sky, and took this as a sign of divine support in his cause from the Christian God.

Q *What was the colossus of Rhodes?*

A Rhodes is a Greek island in the Aegean Sea. It was there that, in the second century B.C., the sculptor Chares constructed a gigantic statue of the god Helios. The statue was made of bronze reinforced by iron and stood 105 feet high. It was located next to the harbor entrance of Rhodes, and drawings we've seen show the statue straddling the harbor entrance with ships sailing between the legs. This is preposterous. The statue was located to one side of the harbor entrance. It stood for 56 years before an earthquake snapped it off at the knees. The remains lay there for 400 years before the bronze was carted off on the backs of 900 camels. It must have been a magnificent sight, and is classified as one of the seven wonders of the ancient world. By the way, of the seven wonders, only the pyramids of Egypt still survive.

Q *Considering the violence of the Mount Vesuvius eruption in 79 A.D., why is the buried city of Pompeii so well preserved?*

A Vesuvius is a composite volcano, one that alternates violent eruptions with relatively quiet ones featuring mostly lava upwellings. But during the eruption you mention, most of the force was directed upward harmlessly. The contact of the intense heat and particulate lava (called tephra) with the atmosphere caused torrents of rain that intermingled with the ash and cinders. This created huge mud flows that overwhelmed the city of Pompeii as well as the towns of Herculaneum and Stabiae. These towns were buried under 20 feet or more of volcanic mud, which is a great preservative, sealing out elements of weathering. Even the circles of wine from glasses on counters were preserved faithfully. The popular belief that all the people died there is untrue. There was plenty of warning and most of the population escaped. However, it should be noted that Pompeii was built on a former ancient lava flow. People never learn, it seems.

Q *When did man first start drinking coffee and tea?*

A The origins are rather vague. Both coffee and tea are stimulants owing to the caffeine content. One legend has it that goats were observed munching on coffee berries and then frolicking around in a rather ungoatlike manner. So the goatherd tried it. This was about 1200 years ago in Kaffa, Ethiopia, hence the name. Later, it was learned how to steep coffee into a brew.

Tea apparently goes back much further, and existing records indicate that it was being cultivated as early as 350 B.C. Its origin is in northern China, from which it spread to Japan in the sixth century. The East Indian Company with its clipper ships introduced the oriental brew into England and colonial America between 1600 and 1858. The Tea Tax that England imposed on Americans led to the famous Boston Tea Party, contributory to the American Revolution. To this day, the English drink much more tea than Americans. Can it be that a certain resentment still exists?

Q *Concerning the expression that someone "ought to be boiled in oil," was this ever done?*

A Yes, many times, and worse. A variety of unspeakable tortures have been practiced, extending back into ancient times. Use of such cruelty, usually ending in death, was justified as in the case of extracting a confession from a criminal. Later, torture was used against political enemies or those opposed to established religion as in the case of the infamous Inquisition. And torture was also employed for revenge. The most repugnant application of torture was for "sport," as in the Roman games under such emperors as Nero. A variation of the boiling in oil torture was to suspend a person over a fire and blast him with oil. On one occasion, Nero had 5000 women, men, and children thrown to wild beasts over a 30-day period. While such cruelties no longer occur, torture of enemies by more sophisticated methods such as brainwashing still take place in different parts of the world.

Q *Isn't there a wall in Great Britain that is like the Great Wall of China?*

A Yes. It is neither as long nor as elaborate as the Great Wall, but is nonetheless impressive. This wall is called Hadrian's Wall and is 73 miles long, extending from the Tyne to the Solway, and was supposed to guard the northern frontiers of Rome's British province from hostile tribes to the north in much the same way as the Great Wall functioned. It contained turrets or forts at two-mile intervals along its length with a protective flat-bottomed ditch. Like the Great Wall, it failed in its purpose, being breached by attackers and destroyed twice in the second and third centuries.

Hadrian, the Roman emperor who ordered the construction of the wall, was an interesting man. In his 20 years of rule, he spent 12 years away from Rome traveling through his provinces. He was a soldier-scholar of great intelligence, but spent the last several years of his reign, despite his power, a sad and lonely man.

Q *The dragon ritual is associated with the Chinese New Year. Do other cultures incorporate dragons?*

A Dragons are very common in art and mythology. Think of Puff the Magic Dragon, Smaug of the Tolkien fantasies, and that fire-breather slain by St. George. Furthermore, the dragon theme is more than 4000 years old, having been found in ancient Egyptian and Sumerian art. Dragon folklore throughout the ages probably had different meanings in different societies. In Europe, for instance, the dragon was a symbol of evil for contrast with the Christian religion.

Dragons sometimes functioned as gods since they were believed to have control over life and death. As mentioned in your question, the dancing Chinese dragon is most familiar to us today—a modern survivor of the time when hundreds of different dragons were recognized in Chinese culture.

Q *Where did Grimm get his ideas for the fairy tales I read as a child?*

A Actually, there were two Grimms—brothers—named Jacob and Wilhelm, who wrote the fairy tales more than 100 years ago. *Grimm's Fairy Tales* were a by-product of their professional interest in language, grammar, and history. From the time both were university students, they were fascinated by folklore and mythologies that had been handed down by word of mouth in Germany. They avidly collected these tales from whatever source and fashioned them into readable stories. The stories encompass the magical lore of many lands, including elves, giants, and animals transformed into people and vice versa. The Grimm brothers would have been famous even if they had not written the fairy tales. In the early 19th century, they undertook to compile a great dictionary of the German language, a task so difficult that it was not completed until 1960 by other scholars. Both brothers worked closely together all their lives and are buried side-by-side in Berlin.

Q *What was the object so searched for by knights in the Middle Ages?*

A The Holy Grail. There are many legends about this that extend back many centuries. The Grail, in the most popular version, was the cup that Christ drank from at the Last Supper. It could also be interpreted to be a dish. As such, it would be a priceless relic if it could be found, thus the stories of the knights of King Arthur's Round Table and their adventures in trying to find it. Doubtless there were containers of food and drink that Christ used at the Last Supper, and for that matter, any vessel He used during His lifetime and which was authenticated as such would be priceless. It is most probable that such objects have long been lost or destroyed. It should be noted that tales of sacred cups and dishes can be found in other cultures, predating Christianity and seemingly associated with early ideas of immortality.

Q *Was there actually a group of knights who wore white robes with a large red cross?*

A Yes they did exist and were, in the 13th century, one of the richest and most powerful organizations in Europe and elsewhere. They called themselves the Knights Templars, and were formed to free the Holy Land from the Moslems. Thus, they were an important part of the Crusades. Members had to have special qualifications to join, and were inducted with secret initiations. They became so wealthy over the years that they drew the envy of Philip the Fair, King of France, who was in need of money. He set out to destroy the Knights Templars. On trumped-up charges of religious heresy, he ordered the arrest of all knights in France and seized their houses and other property. He persuaded the English king to do likewise. Pope Clement agreed to all of this, being a weak pope who owed Philip some favors. Sadly and unjustly, many of the knights were subjected to torture and burned at the stake. In 1313, the Knights Templars officially ceased to exist.

Q *What was the world's greatest archaeological fraud?*

A Some would say it was the famous Piltdown forgery. Piltdown, the phony fossil, was "discovered" in a gravel formation in southern England in 1912. The unknown person or persons who combined an ape jaw with skull fragments of a modern man did the job with great skill, so much so that scientists argued over Piltdown for about 40 years.

The forger had filed the teeth in the ape jaw to give them the appearance of a human wear pattern. Many anatomists of the day accepted Piltdown as a primitive form of man but chemical investigations in the 1950s showed the cranium and jaw to be of different ages. Today our information about fossil forms is more complete and scientists might not be so easily fooled.

Q *The book* America B.C., *by Barry Fell, claims that Celts, Egyptians, and others were present in North*

America long before Columbus or even the Vikings. Is this true?

A It is highly unlikely because no evidence has been found to support this idea. What Fell offers as evidence is not evidence. What he thinks are stone carvings of Celtic origin are simply natural geologic features, which often may resemble outwardly something of human manufacture. Writings in stone turn out to be scratches made by plows scraping stones in New England during the last century. Because Fell has no training in either geology or linguistics as far as we know, his opinion carries no more weight than that of the average person. Yet, because his book makes sensationalist claims, people buy it. As with Von Däniken, of Ancient Astronauts fame, the books sell well to the gullible, more so than a scholarly, well-documented book that would show such assertions to have no basis in fact.

Q *Is there evidence of the Vikings being in North America hundreds of years before Columbus? How would the Vikings have navigated at that time since they didn't have magnetic compasses?*

A No one can say when or where the mariner's compass was invented but you are probably right about the Vikings since they seem to have already crossed the North Atlantic before the compass was known in Scandinavia.

Even earlier, other maritime civilizations such as the Minoans and Phoenicians made ocean voyages out of the sight of land without a compass. Early seamen probably used a knowledge of directions from sunrise and sunset and prevailing wind patterns.

However, the Vikings may have had another navigation aid. Scandinavian sagas refer to "sun stones." Some scientists believe that these were minerals with polarizing properties. These crystalline stones turn color when pointed in the direction of the sun, even when the sun is obscured, as it often was where Vikings sailed.

5
THE SUPERNATURAL

Witches, Demons, Ghosts, and Other Monstrous Things

Q *Do the words "paranormal" and "supernatural" have the same meaning?*

A If we take the words "natural" and "normal" to have pretty much the same meaning, and add the prefixes "para" and "super," then nothing is different. These prefixes suggest something above or in addition to the noun. Yet current usage separates these terms somewhat. "Supernatural" has always implied such mysterious things as ghosts, haunted houses, and "things that go bump in the night." These sorts of things are generally beyond science's present capability to study them. On the other hand, the word "paranormal" has frequently been used to describe such things as mental telepathy, clairvoyance, psychokinesis (the ability of the mind to move objects), and precognition (knowing things in advance). These latter categories are indeed being studied using scientific methods. Although the distinction is a little hazy, you might think of the paranormal as things able to be studied now by science, and the supernatural as being somewhat outside the realm of science.

Q *Would you agree or disagree that magic and science are closely related?*

A We would probably agree, but only in a certain context. If you could (magically!) transport yourself back to the Middle Ages, and show people of that time a tape recorder or a cigarette lighter, they would doubtless call it magic. You would be proclaimed a magician or maybe a god. Less fortunate, you would be thought of as a demon and be burned at the stake.

The point is, whatever is not understood, or does not seem to conform to the laws of nature, is often viewed as some kind of "magic." All cultures through the ages evolved their own superstitions and magical practices when confronted by the unknown. As knowledge advances, much of what was formerly "magic" becomes a scientific truth. However, the entire question is a complex one and our answer is certainly not the final word.

Q *What exactly is magic, and does it really work?*

A To define magic is like trying to nail a custard pie to the wall. There are many definitions, many practices, and for different purposes. The use of magic seems to be universal, being practiced in all cultures past and present. We think there are some common threads here. Most cultures employ objects or other kinds of symbolism together with incantations. These are used in various ways in an attempt either to effect some degree of protection against real or imagined evil, or to punish or harm some enemy.

Does it really work? It seems to be a matter of belief. If you believe that a witch doctor can put some kind of "hex" on you that is harmful, even causing death, it could happen. There are documented cases in Africa. On the other hand, if you scoff at something like a hex, and one is invoked, can you be harmed? That remains one of those mysteries we haven't solved!

Q *Do fringe ideas and extraordinary claims in any way aid the scientist?*

A A whole book could be written on this since the subject area is so vast, but in general the answer is yes. Take, for example, geographical myths. It was once widely believed that there was an undiscovered Northwest Passage to the Orient and also, in the South Pacific, a great southern continent. Neither of these exists, but the belief that they were there somewhere, yet to be discovered, influenced geographers and explorers to search and discover the true nature of the physical world.

Other esoteric claims such as Bigfoot, pyramid power, and UFOs might someday advance modern science. The borderlines of scientific knowledge do change.

Q *What does that magic word "abracadabra" mean?*

A This word has been in use for nearly 2000 years, having originated in Roman times. At that time, people believed in a god called Abraxas who could help shield a person from evil if this god's name were inscribed on stones or jewelry and worn on the person. Use of the word "abracadabra" was thought to cure illnesses, especially fevers. The word was uttered often during the Middle Ages at the time of the bubonic plague in Europe to ward off this deadly disease. Nowadays, stage magicians use the word in their illusions of making things vanish or appear or other tricks. Thus, the word has lost its original meaning. Nevertheless, the next time you get sick, say "abracadabra."

Q *What does a witch do in the performance of witchcraft?*

A A witch performs acts to cause evil to visit an enemy or acts to achieve certain other objectives. Some of these objectives are not necessarily evil. For example, a man may wish to have a specific woman love him. From Paul Christian's book, *The History and Practice of Magic*, consider the following advice: "Take a dove's heart, a swallow's womb, a sparrow's liver and a hare's kidney, and after having dried them, reduce the whole to a very fine powder, to which you will add an equal quantity of your own blood and leave the mixture to dry. If you make the person you desire eat it, she will not be able to resist you for long." There is no scientific basis to suggest that such a concoction will work, and there might also be some objections by the sparrows, doves, and other animals that become party to this plot.

Q *Did most of the executions for witchcraft take place in England?*

A By no means. It was a madness that engulfed much of Europe as the Middle Ages came to an end. Some of the most horrible stories on record come from Germany. In a 13-year period in the state of Bamberg alone, 300 unfortunates were put to death, earning that region the title of the "shrine of horror." While there may have been some motivation to save the souls of these people, other motives were involved as well. The property of those executed was confiscated and wood cutters who sold wood for the burnings at the stake were enthusiastic accusers. It seemed for a time no one was safe. If a person expressed the slightest hint that another was a witch, they themselves often were arrested and put to the torture. Under severe torture, most people will confess to anything. The Roman Catholic Church often is singled out as responsible for all these killings, but it should be noted that the Protestants also burned people at the stake for witchcraft.

Q *How can the fact that many sick people are cured by witchcraft be explained?*

A There are logical reasons for such recoveries. The human body is certainly very durable and often recovers using its own natural defenses, as it did for thousands of years before modern medical science. Most witch doctors are well aware of the curative powers of certain herbs and other concoctions, and also the value of massage and other physical therapy. Perhaps most important is the psychological factor: the most dramatic recoveries are among those with the greatest conviction and belief that witchcraft does work. Thus, when a healer acts to cure a person, the fear of illness, which has often caused the illness, is dissolved and the person is on the road to recovery. Be that as it may, we still recommend that you see your physician when you are sick.

Q *Where did the idea originate that sticking pins in a doll by a witch could harm the person the doll represents?*

A This belief cuts across many cultures and seems rooted in childhood perceptions. To the child, the doll was real and could be scolded or praised or punished. At the adult magical level, the doll is made of wax, clay, or other material, and made to resemble the living person as closely as possible. Incorporation of blood, hair, nail clippings, or other objects related to the victim makes the magic even more powerful. Pins or thorns thrust into any part of the doll will inflict pain or destruction on the corresponding anatomy of the real person. To the anthropologist, this type of witchcraft is part of what is known as imitative magic. In modern garb, athletic teams choose names like Tigers or Bulls, suggestive of their triumph over an opponent by using the strength of the animal involved. To some, the doll serves as a focus for the hate of the witch and indeed, in some societies where this form of magic is believed to succeed, it has been known to have worked, even causing death. It may also be used to make a man love a woman rather than inflict physical harm.

Q *When was the last time a witch was burned at the stake?*

A Witchcraft has always been regarded as a practice of black (and thus evil) magic wherein persons enter a pact with the Devil to gain the power to carry out evil purposes. During the 1600s and 1700s, there was almost a paranoia against witches not only in the United States but also in Europe. People were hunted down and tortured to obtain confessions that they were witches. Under severe torture, most "confessed." In Europe, the last official burning at the stake occurred in Scotland in 1727. In colonial America, several people were put to death in 1692 in Salem, Massachusetts. After these times, official government-sponsored witch-hunts declined. Yet as late as the 1950s, two people were hanged in Mexico who were thought to have been witches. Thus, even today, somewhere in the world, people are still probably accused of being witches, and perhaps losing their lives as a result.

Q *How did the belief in witches actually begin in Salem?*

A In general, ideas and beliefs about witchcraft among the people of Salem were influenced by European attitudes toward Satanism. The Salem witchcraft episode seemed to start in the late 17th century with the arrival of a new pastor, Samuel Parris, together with his wife, daughter, and niece. They were accompanied by two young slaves, one a young West Indian girl named Tituba whose father had been a tribal witch doctor. In what might be called an early American slumber party, a group of young girls began meeting with Tituba in the evenings in order to practice "magic." History records that shortly after these meetings, the girls exhibited a pattern of strange behavior, having hysterical fits and convulsions. Doctors could find no physical cause and concluded that the children were bewitched. This belief grew and eventually over 200 persons were thought to be possessed by evil spirits. From the Salem Village witchcraft trials, nineteen people were put to death in 1692, not including two dogs that were hanged as witches.

Q *Witches are supposed to have "familiars." Are they demons?*

A They were thought to be demons or imps by the witch-hunters. They took the form of a small animal such as a mouse, rabbit, frog, or toad, but most often it was a cat. The familiar was believed to have been given or sold to the witch by the Devil to be used to assist or carry out the witch's work of casting evil spells upon innocent people. The idea of a witch's familiar seems to have originated in England around the 15th century because it is noted that in stories of witches of other lands, familiars are hardly mentioned. Perhaps certain elderly women in their loneliness became strongly attached to a pet cat or other animal. When they were accused of witchcraft, the cat got the blame too. Nowadays, it seems hardly possible that such superstitions led to the torture and execution of numerous women, and some men as well.

Q *My grandmother claims that her cousin was killed by the evil eye in Italy. What exactly is this evil eye?*

A The evil eye is one of the most widely believed ideas in the world. In one form or another, this belief is found in much of Africa, the Mediterranean, and Europe, although it is not found much in the Pacific. The basic idea is that some people (or animals) can cause harm simply by glancing at other people or property. Sometimes the evil eye is cast maliciously upon an enemy, but other people can have the evil eye involuntarily like a sickness. Certain people are often avoided for fear they have the evil eye, such as one-eyed or cross-eyed persons, hunchbacks, or others with some deformity.

The origins of this belief are not known precisely. As

is well known, the eyes can reveal emotions and hidden feelings, and are called the "window of the soul." We might infer that cases of anger or hatred projected from the eyes to another individual followed by some coincidental catastrophe gave rise to the belief.

Q *Was Cagliostro an infamous and evil magician?*

A Cagliostro was probably not a bad person; condemned mostly because of jealousy and suspicion. As we read it, he was a man of active intellect interested in the occult and the supernatural at a point in history when such things were frowned upon by the church. Born in 1743 in Sicily, Cagliostro grew to adulthood gaining skill in such matters as hypnosis and grasping the elements of psychology such that he affected some marvelous cures of sick people and gained wide fame. He also dabbled in alchemy, trying to change lead to gold (although he was rich and needed no money), and other occult matters. Thus, he was held to be a magician. This was dangerous. Unjustly accused of the theft of a royal necklace in France, he went to Rome, where he fell into the hands of the Inquisition. He languished in prison until 1795 when, according to some reports, he was strangled by his jailor at the age of 52. To some he was a charlatan, to others a showman. Cagliostro was not evil, but rather sought to help others. He lived at the wrong time in history.

Q *Can a person under hypnosis be forced to do something against his will?*

A Probably not. While people under hypnosis are in a highly suggestive state, they will not do anything that violates their moral precepts. For example, if a young lady under hypnosis is told to undress, she will come out of hypnosis and be quite angry, although she may not understand what she is angry about. However, if she is a stripteaser, she might undress without hesitation. A person under hypnosis is able to reason only deductively. If the hypnotist suggests that you are a dog, you will act out what a dog does, that is, getting down on all fours, barking, and panting. If, however, it is suggested that you bark, pant, and have four legs, the hypnotized person will not draw the conclusion that he or she is a dog. Hypnotizing another person should not be seen as a parlor game. You need training in this field before attempting it.

Q *If someone puts a curse on me, what can I do about it?*

A It might be wise to do nothing. A curse seems to have an effect only if the recipient of the curse really believes it will. In some societies where such belief is universal, a person who has been cursed is avoided by everyone. He becomes a social leper, an outcast. Psychologically, the effect is devastating and often lethal. Yet, most curses are not invoked to kill, but to punish. A man may see his cattle die one by one, and know he is under a curse. As late as the middle of this century, a man in Arizona killed a person he thought had placed a curse upon his wife. In ancient Egypt, curses were made against anyone who desecrated the grave of a pharaoh. Thus, a curse was a form of protection. In effect, a curse is only as powerful as the depth of belief in its power, and clearly has psychological roots. If you take our advice, however, try to avoid being cursed.

Q *How did Halloween originate?*

A It was a pagan celebration among the Celts and other peoples in northern Europe well over 2000 years ago. It was regarded as the time of the coming of winter and it was believed that the dead returned to their former homes looking for food and warmth. Thus, it came to be that food or other offerings were placed on doorsteps for the dead. Today's trick-or-treaters represent these ghosts looking for a handout. Demons and witches were also believed to roam the darkness on this night, and real witches and their covens continued to celebrate it into modern times. About 1300 years ago, it was made a Christian feast day honoring saints and martyrs and called Allhallows or All Saints' day. The practice of children going from door to door for treats is a fairly recent custom, one that seems to be threatened by the appearance of poisoned candy in some of the offerings.

Q *Does the water from Lourdes have magic curative power?*

A Yes and no. Bernadette of Lourdes was a child of 14 in 1858 when she claimed to have received several visions of the Virgin Mary at a grotto there and found a spring that has been flowing since. Today, the town receives over two million visitors each year, many of them hoping to be cured of ailments. Chemical analysis of the springwater shows it is good drinkable water, but without any special properties. Vials of this water are sold with the idea it will cure diseases and other infirmities. The church itself is rigorous in its judgment of claimed healings at Lourdes. And there are few such cures, considering the millions of people who have been there. On the other hand, the powerful effect on the mind and spirit—the psychology of it—may well have a beneficial effect for the afflicted. Bernadette herself lived only to age 35, dying in a convent in great pain.

Q *How can a woman be burned to ashes while the chair in which she was sitting remains unscorched?*

A Such reports defy all logic of science and yet there are documented reports of spontaneous human combustion that have been investigated by the police and even the FBI. A common feature is that the torso is consumed, including the bones, while the extremities remain at least partly unburned. Most of the victims are women. Also, the surrounding area, including floors, walls, and furniture, is not burned or even scorched. We have not heard of the case you cite and we consider it an exaggeration. True, many of these cases are hoaxes of some kind. For those that aren't, the only scientific explanations we know of are that the victims were alcoholic and this encouraged rapid burning of body fats. Another explanation is the ignition of gastric gases. But how does ignition take place? It also requires intense heat to reduce bone to ashes. One eyewitness said the woman he was with burst into flames as though the fire emerged from within head and chest. This is one phenomenon for which we have no pat answer.

Q *Is levitation possible?*

A There are no known laws of physics or any other science that allow this possibility. Certainly, astronauts in outer space can float around weightlessly, but not here on the earth. Many reports of levitation, usually a few inches off the ground, come from religious sources. It is claimed that in a state of religious ecstasy, certain holy persons have levitated in the presence of many witnesses. Such reports are fairly numerous over many centuries, although the numbers diminish in modern times.

D. D. Home, a medium in the 19th century, was the most famous air traveler. In 1868, he supposedly floated out a third-story window, 70 feet above the ground, and returned by way of a different window. Verification of such happenings is difficult: one eyewitness said, "It was too dark to see how his body was supported." Home was never discredited. There is always the chance he was using magician's tricks. On the other hand, we do not know with certainty.

Q *How did the idea of the Devil get started?*

A The idea of the Devil today is associated with early Christianity as there are several references to the Devil in the Bible. However, the idea may go back much farther to the time of cave-dwelling primitive man. Cave drawings thousands of years old depict a man wearing animal skins and a pair of horns. These drawings may be the forerunner of the Horned God that represents the forces of nature and is the earliest known deity. This god symbolized both good and evil. As civilizations developed, the notions of good and evil became more crystallized, leading to separate personifications for good and evil. The idea today that the Devil has horns, hooves, and a tail may be traced back to the Horned God. The Devil has been called by many names because almost all societies and cultures symbolized evil in some way.

Q *Why is the goat so often associated with the Devil, or a disguise Satan uses?*

A The origin of this idea is obscure. Goats were worshipped in Egypt more than 2500 years ago, but not as an evil entity. During the many witch trials of the Middle Ages, accused witches confessed to confronting the Devil taking the form of a goat, and hearing the goat speak like a man. There is some suspicion that sometimes the goat was actually a man clothed in animal skins. Adding to the idea of the Devil as a goat is its reputation as a rather smelly and repugnant animal, and its often black color. This is an undeserved rap against the goat, which through many centuries has been a very useful animal for mankind. Early man used the milk, the skins for leather and clothing, and the meat for food. And we are still using the domesticated goat. Goats' milk is very good and more digestible for infants than cows' milk. Some goats can produce more than 6000 pounds of milk in a year and the meat has a delicate flavor similar to lamb. Yet the goat has had a devil of a time gaining the general admiration of men.

Q *Why is a pitchfork associated with the Devil?*

A The exact origin is unknown. We might surmise that it started out as a male symbol in the same way that the broomstick was the symbol of a woman; that is, in olden times the tool used to carry out domestic chores. The pitchfork represented the man, working in the fields. Somewhere along the line, these two symbols became associated with witches and demons, perhaps because they were carried to nightly gatherings of alleged witches and warlocks. A more obvious use for a pitchfork by the

Devil in the realm of hell would be to prod and torment the condemned souls there.

Q *Why are certain musical instruments (flute, violin) associated with the Devil?*

A Music in various forms and instrumentation has been associated with nearly all religions for thousands of years. Music (and also dancing in some cases) has the power to evoke emotion and ecstasy, bringing the participant into a perceived spiritual state. This is all well and good, but as various religions developed, some sects were thought to be more in league with the Devil and their ritual music regarded with suspicion. Thus, the violin and flute (associated with the pagan god Pan) were viewed adversely. Indeed, at one time, quite an argument took place over the use of the organ in Christian ceremonies. Drums are another instrument generally associated with non-Christian ritual. Before the Christian era, prophets used drums, singing, and dancing to induce the physical and mental state to be able to see into the future. The state induced was more likely one of self-hypnosis.

Q *What is the truth about the mysterious footprints made in the snow one night in Great Britian that were seen for many miles and thought to be the footprints of the Devil?*

A There were many in Devon who woke up that morning in February 1855 and believed the Devil had strode through their towns and villages. There were prints in the snow of cloven hooves similar to those of a donkey, perhaps three or four inches in diameter. These prints were spaced about eight inches apart and, incredibly, trended in a straight line through streets and backyards, over fences, and even crossing snow-covered roofs of houses. The people of the region assumed at first it was some kind of animal or perhaps several that had romped through their area during the night. But what donkey or other creature could move over eight-foot walls without breaking stride? These unusual tracks extended over 100 miles and even crossed two miles of open water in an estuary. It would seem nothing of this world, and that indeed may be the answer. Our best guess is that the county of Devon was visited that night by a meteor shower, with small fragments impacting silently and making impact markings that, while resembling prints of a hoofed animal, also resemble some impact craters on the moon. At that time, most people did not believe the assertions of some scientists that stones could fall from the sky.

Q *Was there a real Faust who made a pact with the Devil?*

A George Faust was a real person who lived during the 16th century, dying about 1540. According to most accounts, he was a good-for-nothing braggart who claimed magical power and performed tricks for audiences around Germany. Among his claims was that he had a close relationship with the Devil, and referred to the Devil as his "crony." This was probably pure baloney to impress naive audiences. No doubt Faust would have faded fast into history but for the fact that Protestant churches seized upon Faust (after his death) as an example of what can happen—eternal damnation—if a person sought too much knowledge. Books and plays were written on this theme, especially by Christopher Marlowe and later Goethe. However, Goethe modified the theme somewhat by showing that even an evil person such as Faust might be forgiven by God. It is interesting that Faust, who himself contributed little to humanity, became an inspiration for a body of outstanding literature.

Q *Why do drawings of saints show a yellow halo around the heads?*

A This is the aura, thought by some to surround all people, but especially holy people with great spiritual power. Such representations are known from art dating from the fifth century and perhaps earlier. Sometimes the aura envelops the whole body and may be other colors such as green and violet. This is not strictly of Christian origin, as holy persons such as Moslems are depicted as being surrounded by flames rather than a glow. Also, even those thought to be evil (e.g., the Devil) are sometimes shown as enveloped in a glow. In a real sense, each of us has an aura—one of heat—that surrounds us. That is how snakes can seek out and kill small mammals in the dark. While some people claim to be able to see this aura, there is no scientific evidence that this is possible.

Q *Just what are demons? Is that what the Amityville Horror was? Are they like ghosts?*

A The word "demon" is a general term used to describe an evil spirit, although in some cultures a demon could be a good spirit. Indeed, they have been regarded as the ghosts of humans seeking revenge for alleged wrongs, and able to take on bodily forms such as werewolves, harpies, and vampires. Demons of nonhuman origin have been depicted as followers of the Devil, who himself is a "superdemon." Demons have the reputation of lurking at night in forests and byways to pounce upon and harass travelers. Mentally disturbed persons in the Middle Ages (and even today) were thought to be possessed by a demon and required exorcism. If it did in fact exist, the so-called Amityville Horror would be considered a demon. In that case, though, there seems to be considerable exaggeration and embellishment. Nobody we know has captured and dissected a demon, so the concept is without scientific foundation. Nonetheless, the subject of demons makes for some good, hair-raising stories to be told on dark, stormy nights.

Q *Are there genuine cases of demonic possession such as shown in the movie* The Exorcist?

A The movie illustrates very well the symptoms of possession that apparently are manifested in such reported cases. These include physical effects such as vomiting and frenzy, and mental effects such as personality change, use of filthy language, and so on. People have believed in demonic possession for many centuries. Why is "God bless you" said after someone sneezes? It was believed that the soul was expelled by a sneeze and at that moment a demon could enter and take control of the body unless a blessing was given. Most of the cases of possession we have heard about involve individuals who are deeply religious and believe strongly in demons. For this reason, some cases might be due to strong autosuggestion. Psychologists have noted that symptoms of possession are very similar to those of persons who are mentally ill or disturbed. Exorcism parallels psychiatric treatment of mental patients. Many cases of possession took place long ago and are poorly documented. These cases can't be evaluated properly.

Q *Is the Loch Ness Monster the only monster reported in Scotland?*

A The Loch Ness Monster is far from the only monster reported. It seems every lake in Scotland has a resident monster. Examples include the monsters seen at Loch Lomond and Lochfyne, and one from Loch Awe resembling a giant eel and described as "big as a horse with incredible length." If you look anywhere in the world where there is a sizable lake, there is also a monster: Ireland, Canada, Australia, Mexico, and so on. There are monsters in most lakes in the United States such as the one in Lake Walker, Nevada, that (so it is said) is a man-eater, but eats only white men by special agreement with the Indians. Most monsters are not bad and have affectionate nicknames such as Nessie in Loch Ness and Champy in Lake Champlain, New York. Such monsters are created by mistaken identification or imagination, to increase the tourist trade, or made up just for fun. In 1934, Sir Arthur Keither said, "The only kind of being whose existence is testified to by scores of witnesses and which never reaches the dissecting table, belongs to the world of spirits. . . . I have come to the conclusion that the existence or non-existence of the Loch Ness Monster is a problem not for zoologists but for psychologists."

Q *If there actually is a Loch Ness Monster, why hasn't it been caught?*

A There is enough good eyewitness testimony to conclude that there is something there, but what? A favorite theory is that it is a plesiosaur, a sea-going reptile that became extinct 70 million years ago along with the dinosaurs. Scientists don't think so. The plesiosaur lived in shallow salt water and was an air-breather. The Loch is deep fresh water. Sightings of the monster are too infrequent for an air-breathing animal (only one sighting for every 350 hours of observation). Most sightings have been of snakelike humps breaking the surface of the water. It could be an unusually large eel. There are many eels in the Loch. They like deep, dark water and are not air-breathers.

To catch the monster is a tall order. The water is stained dark brown by peat so visibility is poor. Water depths reach 950 feet. The monster has a great hiding place, even if a big effort is made to catch it. That effort hasn't happened yet to our knowledge.

Q *What of Bigfoot? If it exists, why haven't any of its bones been found?*

A You've put your finger on a crucial point in the Bigfoot debate. Like Nessie in Scotland, the evidence for Bigfoot tends to be insubstantial and disconnected. Alleged sightings, footprints, and fuzzy photographs are what scientists call soft evidence. If we could find bones, they would be the hard proof needed to convince the skeptics.

Those who believe in Bigfoot say that bones don't survive. How often, they ask, do you ever see deer or bear bones? Nature, in the form of mice, weasels, squirrels, and chipmunks, sees to it that even the bones don't last. On the other hand, nature does have accidental ways of burying and preserving bones. Many bones of fossil remains are found from various stages of evolution but none from the manlike creature called Bigfoot. This may be because the animal doesn't exist.

Q *What is the origin of the werewolf legend?*

A It is anybody's guess, as these stories go back to the time of ancient Greeks, perhaps earlier. The best known legend is about a man named Lycaeon who sacrificed a child's flesh to the god Zeus. Aparently Zeus didn't like this and punished the man by turning him into a wolf. Based on this, the werewolf condition came to be known as lycanthropy. The word "werewolf" means "man-wolf." The origin may also date back to times when Scandinavian warriors donned the skins of feared animals, such as wolves, before they attacked their enemies.

Psychiatrists have treated people who imagine they can be transformed into a wolf under the influence of the full moon and then return to normal after a night of prowling for human flesh. The whole idea is known in many countries and under many guises. The true origin may never be known precisely.

Q *Has any person ever been raised by wolves or other animals?*

A It is really doubtful. The classic case, which is founded in myth, is that of Romulus and Remus, twins who were raised and nurtured by a she-wolf. Romulus and Remus were involved in the founding of Rome, and indeed Italian postage stamps portray the twins suckling from the she-wolf. Over the years there have been many stories of children raised in the wilds by wolves, bears, or other animals. These children moved on all fours, ate raw meat, and could only make the sounds of the animals that raised them. A specific case was that of the wild boy of Aveyron, France; in 1797, three hunters found him climbing trees and eating acorns and snarling like a beast. He was brought to Paris and exhibited in a cage, tragically enough. Most of such stories are without solid documentation. Those that seem authentic probably involve children who suffer from retardation or other mental deficiency. From the animals' point of view, we know of no wolf or other animal willing to—humanely—take a human child under its wing and raise it.

Q *What is the meaning of the totem poles carved by Indians in the Northwest?*

A Surprisingly, this is a complicated question. The use of animal or plant symbolism in tribes or groups has been a universal practice. Through the ages, man has had a close relationship with the animal world. Man admired many of the physical traits of the animals and wished for the "agility of a cat" or the "keen eye of an eagle." Such traits today may not seem important, but in the past they were equated with survival. Some Indians believed that their ancestors could change their shape into that of an animal by magic and this took the form of ancestor worship, as a totem, within specific groups. The function of the totem, in many cases, is as a sign of group identity, a religious symbol, and a form of protection. The word "totem," derived from the Chippewa language, originally was "*ototeman*," meaning a close relationship. Don't think that modern man has gotten away from this idea of totems. We still have the custom, for example, of calling a football team the "tigers" or the "panthers," hoping, we guess, that the team involved would live up to its name in defeating its opponents. Many things don't change.

Q *Are there vampire bats that suck out the blood of the living? Was there a Dracula?*

Q There are vampire bats that drink blood. However, they do not "suck" out blood. They make a wound with sharp teeth, secrete an anticoagulant, and lap up blood as the wound bleeds. Such bats are found in South America, the most common one being called *Desmodus*. He's all of three inches long. The victim's greatest danger is from rabies, not loss of blood.

Dracula is largely a Slavic legend. Perhaps its origins can be found in the few deranged persons who occasionally drank blood for any number of reasons. Blood symbolizes life and vitality. The concept of an "undead" person emerging from a coffin at dusk to feed upon the blood of the living (à la Hollywood) has no scientific basis, but it does make a tantalizing story to tell at bedtime.

Q *Why was garlic thought to repel vampires?*

A It isn't surprising that garlic was suggested because garlic cloves have been thought to cure just about everything except bad breath. It was said that if you carried around a piece of garlic, you'd never get sick. In addition, it was a powerful shield against evil spirits of various sorts and demons, including the vampire. Oddly enough, the plant was believed to have sprung forth from the footprint of the Devil when he was cast out of heaven. If so, the Devil landed in central Asia, where garlic seems to have originated. In the United States, it is grown mostly in California. We recommend garlic as a flavoring, but have serious doubts about its powers in most other respects. Considering its "fragrance," it is amusing that garlic is a member of the lily family.

Q *Why are vampires supposed to cast no reflection in a mirror?*

A This concept goes back farther than Dracula. In primitive times, one's reflection in a mirror, a quiet pool of water, or any shiny surface was thought to be one's soul or life force. In some societies, it is considered bad luck to see one's reflection in a pool of water for fear that some supernatural beast will steal it and thus take your life. Vampires were regarded as having no soul, and therefore could not cast a reflection. The belief that bad luck ensues from breaking a mirror is similarly based, because if the mirror is broken while you are looking into it, you may lose your life. There is no scientific basis for these ideas.

Q *Both the bat and the owl are creatures of the night. Why is the bat regarded as evil and the owl a symbol of wisdom?*

A Your assumption is not entirely correct. There are many cultures, perhaps the majority, that view the owl as an omen of evil and death equally with the bat. In other places, especially where the owl preys upon rodents and other animals that eat crops, the owl is thought of as good. Unfortunately for the owl, some cultures nail an owl to the barn door to prevent hail and lightning from striking during storms. The owl's reputation for wisdom seems apparent. They look like they are thinking great thoughts. Not so the bat. The bat is indeed an ugly-looking creature whose association with the Devil and all things evil is firmly entrenched. Perhaps this is due in part to some varieties such as the vampire bat indulging in the drinking of blood and sometimes being a carrier of rabies. In contrast, some Oriental cultures consider the bat to be a symbol of long life. And in some places, the blood of the bat is thought to be an aphrodisiac.

Q *Why is the Grim Reaper a symbol of death?*

A The skeleton, perhaps since the earliest beginnings of man, has represented death. Some of the earliest drawings show a skeleton bearing darts with which to destroy those whose time had come. Thus, death was personified generally in this way, with variations from culture to culture. During the Middle Ages, the idea of time was also personified as an old man with flowing mane carrying a long scythe. Man has long recognized the close connection between time and death, resulting in the combined symbol of time and death as the Grim Reaper, the skull peering out from beneath the cowl, carrying the scythe to cut the life stem of the living. Such representations sometimes carry an hourglass. A less grisly symbol is the Christian angel of death—less grisly, but just as final.

Q *What is the "banshee" that wails as a warning of a person's impending death?*

A It is a legend of Irish origin. Somehow the belief arose that a ghostlike creature associated with families of pure Irish blood would foretell with screeches and wails the death of a member of the family. Most legends have some basis in fact. In this case we suspect that a family pet such as a dog or cat did the howling coincident with the death of a family member. If the pet was out of sight, the conjuring of a supernormal being—the banshee—was not that unreasonable an explanation. The word seems to be derived from the Gaelic *bean-sith*, which means fairy woman, a creature that acted as a guardian of certain aristocratic families. The Irish have other charming legends such as the leprechaun, the little old man with the power to disappear mysteriously and who guarded a treasure of gold.

Q *What is a zombie?*

A Pure invention. The zombie is supposed to be a corpse brought back to life by means of voodoo. Belief in zombies is rife in Haiti, but in few other places. The zombie is created by the use of poisons or enchantment, and used as a slave by the sorcerer. In this condition, the zombie has little power of speech and walks with a shuffling motion. If the zombie eats any salt, it is supposed to reawaken and may take vengeance against its master before expiring. The zombie legend likely derives from the fact that retarded people in Haiti behave like zombies are supposed to, and superstitious people have accepted this belief.

Q *How did the idea originate that fairies exist?*

A The belief in such creatures extends back at least many hundreds of years and its origin is a complex matter. One interesting theory is that during the Stone Age, new settlers in a region displaced the original inhabitants, who then retreated to isolated areas and were not seen very often. Because of their greater familiarity with that region, they were held to be elusive and the idea grew that these people possessed magical abilities. Often, the displaced people were of shorter stature than the invaders, and so the concept of fairies as "little people" became widespread. While modern fairy tales portray fairies as good little people, the earliest beliefs show them as sinister and evil, or at least mischievous. An added reason for believing in fairies could have been that when something bad happened, it could be blamed on the fairies rather than on one's own blunders.

Q *Is Hades the same as hell?*

A It has come to mean that in modern times but they were not always considered the same. Hades was actually a person or god in Greek mythology, otherwise known as Pluto, who ruled over the underworld, which was peopled by the dead in a realm of cold and darkness rather than flames and torment. The word "hell" comes from an Anglo-Saxon word meaning to conceal or to cover up, as in the case of a grave. The concept of hell prevails in many cultures, and it is generally thought of as a place of fire and torment, often eternal but not necessarily so. The earliest information we have about ideas of the hereafter is that heaven was the first to be conceptualized. Later, the idea of an infernal region of punishment for the wicked took form when it was perceived that some people regarded as evil were going unpunished in this life.

Q *Are there such things as "graveyards" where elephants go when ready to die?*

A This seems to be a fascinating piece of folklore. The story is found as part of the plot of Tarzan movies with explorers trekking through the jungles in quest of the secret elephant graveyard so they can take away a fortune in ivory tusks.

Elephants are gregarious and travel in herds. The basis for the graveyard legend may be in the discovery of several skeletons closely associated, and resulting from some common disaster, such as drinking poisoned water, or getting stuck in a bog of quicksand. However, many skeletons are found singly in different places, and not necessarily concealed. As long as we are talking about elephants, we might also add that they do not seem to be especially afraid of mice, but they are intelligent animals and can have a good memory.

Q *How did the dragon myths originate?*

A It is easy to imagine people long ago coming across giant skeletons of fossil dinosaurs, many of which were indeed quite dragonlike. From there, the mythology of the dragon grew. Few cultures have no dragon legends. The beast is often depicted as being a scaly, snakelike creature with wings and sometimes two heads. It lives in deep caves guarding treasures and breathes fire and smoke when angered. The dragon has become mostly a symbol of evil and may personify the Devil. Stories such as St. George killing the dragon may represent the triumph of good over evil. However, in the Orient the dragon is regarded as generally a fine fellow and not evil at all, although when provoked, Chinese dragons have the power to eclipse the sun. Dragons have a reputation for having excellent eyesight and in fact the word is derived from a Greek word, *drakōn,* meaning "sharp-sighted."

Q *Is there any basis for the story about the Minotaur and the labyrinth?*

A Like most myths and legends, it is probably a mixture of truth and fancy. There can be no biological basis for a creature half-man and half-bull devouring boys and girls. In the story, Theseus enters the labyrinth and slays the Minotaur and then finds his way out easily, having unrolled a string or thread as he progressed into the maze. It is a charming but mostly fictitious story, and that's no bull!

On the other hand, mazes or labyrinths were constructed in ancient times and seem to have originated in Egypt. They were associated with temples and rituals and may have been used as burial places. The ancient historian Herodotus described the Egyptian labyrinth at Arsinoe as a building containing 3000 chambers, half of them below ground level. However, there is no archaeological evidence of a labyrinth at Knossos (in Crete) where Theseus's exploits presumably took place.

Q *What is a golem?*

A In Jewish tradition, it would be the equivalent of Frankenstein's monster, a creature brought to life from inorganic materials by means of magic. Fashioned from clay, the golem performed his master's bidding. On the golem's forehead was etched the word "*emeth*" meaning truth. The creature grew bigger each day, reaching great size. It was custom to erase the "e" in "emeth" to alter its meaning to "he who is dead." Then the monster would collapse and be no more. If this was not done, the golem might do harm to its master. *The Sorcerer's Apprentice* by Goethe immortalizes this legend.

Q *Why do people believe in ghosts?*

A It is difficult to pinpoint a universal reason why people would accept the idea that ghosts do exist. Such beliefs, as with most beliefs, serve some sort of valuable social or psychological function. This is not to say ghosts do or do not exist. We cannot be sure. Since a ghost represents an aspect of a deceased person, belief in ghosts may provide reassurance to the living of survival after death. A more recent suggestion that moves closer to science involves the law of conservation of energy. This states that energy can neither be created nor destroyed. That being so, some have claimed that even after death the energy generated from the human body cannot be destroyed and can maintain a cohesiveness (as a ghost) for an indefinite time before dissipating. Perhaps some day we will find an answer.

Q *Is there any proof that there are ghosts capable of moving or throwing objects around in a "haunted" house?*

A There is no solid scientific evidence that we are aware of. Yet many cases have been reported and investigated over the years. Many, if not most, of these cases involve the presence of children. It is likely that many of the "phenomena" observed were simply pranks played by children on their elders for fun or to attract attention to themselves. It is easy, when no one is looking, to throw an object across a room and then quickly play innocent. Yet some—even scientists—are gullible enough to accept these as genuine phenomena. Related to this is the idea of telekinesis, or production of motion in objects without physical means. Most data accumulated about this point to pure fakery. In one case, children were put in a room and told to try to bend spoons using only the power of their minds. When the investigators returned, many spoons were indeed bent. However, by closed circuit TV, the children were seen to have stepped on the spoons. In this area of borderline science, it is best to be skeptical.

Q *How can objects such as ships and planes be seen after their destruction?*

A We suppose you might be thinking of the Flying Dutchman, one of the most famous ghost ship tales. It is true that there are numerous reports of ghost ships, planes, trains, automobiles, and other objects. Probably many if not all of these sightings can be attributed to misidentification, hallucination, mirages, and so on. Such phenomena are not amenable to scientific investigation. However, those cases, and there are several, that seem to repeat themselves might be critically studied. For example, it has been reported that a ghostly sailing ship enters the harbor of a town in Nova Scotia each December, and then burns and sinks before many witnesses. However, until instruments such as cameras have been used to record the events, this report can't be evaluated.

An argument could be made for living beings surviving death as a residual life force, and taking the form of a ghost. The idea of inanimate objects persisting similarly is highly speculative.

Q *Do haunted houses actually exist?*

A Throughout history there have been many thousands of reported instances of allegedly haunted houses. There may be something to this, but it is a difficult area for science to address. There are many natural explanations for supposed hauntings. For example, one such house in England was famous because on wild stormy, rainy nights, a pool of "blood" appeared in an upper bedroom where a man had been murdered. It turned out that the "blood" was a pool of water stained red from leaves that had accumulated between the inner and outer walls. Rainwater trickling down through the walls took on a reddish color as it moved through the leaves and leaked out into the room.

Belief in haunted houses, and thus ghosts, gives us "evidence" of our immortality. Science cannot answer the question posed until one such house has been analyzed and the alleged phenomena scrutinized under strict scientific methods. So far this has not been done satisfactorily.

Q *Is there any truth to the reports that voices of the dead can be picked up on a tape recorder in allegedly haunted places?*

A The method is to set up any ordinary tape recorder and allow it to record under conditions of absolute silence for five or ten minutes and then to replay the tape, listening carefully for ghostly whispers. Occasionally, sounds can indeed be heard that resemble a human voice whispering short statements such as "What's the matter" or "Missed you." The exact origin of these sounds is unknown, but some, perhaps all, are sounds produced by the internal workings of the tape recorder or subtle external noises not consciously recognized. These messages come very swiftly and softly and it can well be debated whether or not they are voices of the departed. It is one area of the unknown where you can investigate and judge for yourself.

Q *Wouldn't the tape recording of the voices of ghosts in a haunted house be proof that ghosts exist?*

A We would suggest being skeptical and looking for a natural explanation if indeed some voices were recorded. The scientific approach is to consider alternative hypotheses and look for the most plausible explanation first. It could be a hoax; they happen every day. But let us say that can be ruled out. In one case, the ghostly voices turned out to be the faint feedback from a previously erased recording. In another case, 26 people were present at the recording session. Especially under the novel conditions of hunting for ghosts in a haunted house, it is difficult to keep that many people from doing a little whispering, perhaps not noticed by others, but picked up by the tape recorder with volume turned way up. Unless rigorous conditions are maintained, little of scientific value can be learned. We have tried these experiments ourselves on several occasions without getting positive results.

Q *Is there any authentic account of ghosts haunting the Tower of London?*

A The Tower of London was constructed many centuries ago and served as the scene of imprisonment and execution of many, including several famous people. It is no wonder that legends of ghosts abound in this old fortress. Of course, many accounts have a natural explanation or are due to overactive imaginations. Repeated sightings have been reported of the ghost of Anne Boleyn, whom Henry VIII had beheaded in 1536. One encounter involved a sentry on guard duty in 1864 who saw a white figure come out of the darkness. When it refused to obey the command to halt, the sentry struck with bayonet, but it passed through the apparition and he fainted. In that condition he was found and accused of sleeping while on duty. At the court-martial, he claimed the "ghost" wore a bonnet but the bonnet was empty. Other witnesses corroborated his story and the court found him not guilty. Whatever it was, it seems something was going on in the Tower of London that night.

Q *Is there any truth to the story that at Fort Niagara, there is a ghost of a headless soldier from Revolutionary times who comes out of the water well at night?*

A None. There is an underground conduit that intersects the well shaft and is big enough for a man to crawl through. According to officials at the Fort, some soldiers played pranks on fellow soldiers on guard near the well by crawling to the well shaft and making mysterious groans and other noises. This would establish the idea of a ghost in the well. It would then be a short step to seeing someone walking near the well on a dark night and thinking it was a ghost. Why headless? The nightly figure was not recognized; therefore, it had no head!

Headless ghosts are always more scary that those that have one. We believe many ghost stories originate similarly.

Q *In these talks some people claim to have had with the dead, how is the hereafter described?*

A There has been a great deal of literature published that supposedly involves conversations with the dead. According to these conversations, the "dead" speak of the "other side" in vague generalities and quasi-philosophical terms. When pinned down, we are told that the hereafter isn't much different than earthly life—there are houses, laws, streets, casual get-togethers for conversation. One dead person named Betty was asked by a medium if there was "electricity and oxygen and bricks and sticks and stones in her world." She answered yes, but that these things were dealt with "not in their obstructed aspects . . . but in their essence." This type of statement is gibberish and sounds like a case of avoiding the real question. We suggest, until we know better, that such talks with the dead are manufactured. Those who are curious about the life beyond need only wait. They'll get there sooner or later like the rest of us.

Q *What is Ouija?*

A There are many who regard this as an amusing parlor game and others who take it quite seriously. The board itself contains all the letters of the alphabet, the numbers one through ten, and the words "yes" and "no." Accompanying this is a pointer or planchette that can move around when one or more persons place their hands on it. Thus, words and numbers can be indicated, purporting to be messages from the beyond. We know of no documented case where such messages contained information not known to the participants. We suppose that on a dark, stormy night, those using a Ouija board might get carried away and unconsciously "help" the pointer to slide around and spell out a message that seems intelligible. We have made determined efforts with the Ouija board without success. The Ouija board gained great popularity some years ago when it was featured in a gripping ghost movie entitled *The Uninvited.* We recommend the movie and also a session with the Ouija board for fun, but do not expect to get any message from your dead aunt.

Q *If a Ouija board has no psychic value, how can the case of Patience Worth be explained?*

A The world-famous case of Patience Worth began about 75 years ago when Pearl Curran, a housewife living in the Midwest, began receiving communications via a Ouija board, starting with, "Many moons ago I lived. Again I come—Patience Worth my name." Gradually the Ouija writing revealed the spirit of a girl who had lived in England during the 17th century.

Over many years, Mrs. Curran served as the channel for a series of literary compositions. There were poems and novels of medieval and Victorian England, judged by experts to be of exceptional literary creativity, allegedly written by the disembodied spirit of Patience Worth. Investigators agreed that the quality and content of writing were far superior to the ordinary powers of Mrs. Curran.

Was there really a Patience Worth communicating

from the spirit world? Or, as some scientists say, was it just an extraordinary case of secondary personality, that is, a creation of Mrs. Curran's unconscious mind?

Q *I am fascinated with the idea of reincarnation. Does science offer any proof that we reincarnate?*

A No. Many people around the world believe in reincarnation, often as a part of their religion. Its roots likely lie in the fact that most people do not want to die. Central to the conviction, for many individuals, that reincarnation takes place is the experience called déjà vu ("already seen"). You come into a strange room and immediately feel you have been there before, in another existence. In fact, you may have been in the same room or a similar one during childhood, but have forgotten. We do not offer this as a pat answer, but only as one possible explanation. We have experienced déjà vu ourselves, and traced it to such origins.

Most reincarnated people we have talked to were queens, kings, princes, dukes, and other prominent personages in the former life. We have met five Napoleons and three Cleopatras. The thing that bothers us is that there doesn't seem to be enough royalty to go around. One consoling thought is inheritance of characteristics. You, as an individual, may think, act, and resemble physically an ancestor who lived 1000 years ago.

Q *Why do so many believe they will live forever after death?*

A Science has no firm answer as to whether or not this will occur after the death of the body. Perhaps so, and maybe not. As to why people believe this, we would note that a primary instinct in all creatures is that of survival. In lower animals, such as dogs and cats, this instinct is physical survival, as we know of no cats that dream of the hereafter. But with man, the rational intellect, regarding the physical death of living creatures as the final end was unsatisfactory. It could be reasoned that people progressing through life often reached their end unrewarded for good works while other evil people escaped without just punishment. Thus, there would have to be a hereafter in which these discrepancies were corrected, if there was a just God. Most cultures believe in a continuance of life after death, and this could be a reflection of the instinct for personal survival. In a sense, belief in immortality is the ultimate fulfillment of the biological instinct for survival.

Q *Do scientists believe in immortality?*

A Many scientists adhere to formal religions of which the idea of immortality is a basic tenet. Other scientists who may not be so affiliated, as well as those who are, would no doubt subscribe to another form of immortality—that is, genetic. All organisms depend upon genes, the cellular elements containing all the information that defines their existence. Genes have the ability to produce exact replicas of themselves. The genes are, in a way, the essence—the germ plasm—of an individual. Through reproduction, this germ plasm remains alive as long as the descendants of the original individuals survive and reproduce. Those genes coding for characteristics that are successful in the evolutionary arena will remain, while those that are not will be weeded out. In a sense, then, if you want to be immortal, have plenty of offspring who grow up healthy enough to reproduce.

Q *Would you speculate as to whether or not man is immortal?*

A Speculate is about all we could do because science itself has little to go on. Either humans continue on in some conscious form after death, or they don't. No unequivocal answer is possible. We are suspicious of claims that a person, say, on an operating table died and came back to life and then reported that they saw a tunnel with light, spoke with long-dead relatives, and even saw God. Appealing as this may be, the temporary stoppage of vital signs can be viewed as "clinical" death rather than true death. According to the conditions science imposes on itself, visions a person experiences at the brink of death cannot be thought of as a brief visitation to the "other side." Another angle to this question is the fact that survival of the organism and its reproduction are the highest priorities of nature. That includes other creatures as well as man. For an imaginative and usually rational creature such as man, the concept of immortality represents the ultimate way to survive, and is expressed in our cultural and religious systems.

Q *Why do people in some countries dance, sing, and otherwise celebrate at a funeral?*

A It is the living, not the dead, who are emotionally wrung by the death of a loved one. They need comfort and an outlet for the grief that they feel. People in other cultures have adopted different methods or rites to alleviate this grief. Singing and dancing at a funeral may be a way of rejoicing that the deceased has passed on to a better life. In the process, the aggrieved also feel better. Even in the United States customs vary. There may be singing and dancing although that is not usual. We know of one case where the deceased instructed in his will that he be cremated and his ashes spread upon the waters of a lake he loved. And then, according to his wishes, his friends, in boats, would toast him with rather strong spirits (no pun intended).

Q *Don't the "dancing coffins" on Barbados constitute proof of the supernatural?*

A It is true that when sealed crypts at two burial sites were opened, the coffins had moved around. Our instinct would be to look for a natural explanation first and think about the supernatural later, if necessary. One crypt had not been opened for a century, and a long-forgotten earthquake may have been the cause.

Coffin movement at the Chase vault is more difficult to explain. It had been opened more than once in recent years, and no earthquake was responsible. The Chase vault was carved from rock to a depth of four feet below ground. It is plausible, according to a suggestion of Herzog's, that the coffins were shifted by the invasion and retreat of groundwater from below. The 600-pound coffins would displace more than 3000 pounds of water and hence would float. The coffins were moved to another site on a hill and have not moved since. When we visited the tomb in 1979, there were small children playing on it and having a great time. No fear of the supernatural there.

UFOs, Superstition, and the Powers of the Mind

Q *If UFOs are visiting the earth from other worlds, just what is their purpose? They have taken their time showing their hand, in my opinion.*

A You have raised a valid point. We have been rather negative about UFOs in the past because most of the sightings turned out to have been meteors, weather balloons, various weather phenomena, and so on. Others can be ascribed to wishful thinking and outright hoaxes. Still others have not been explained. Even if some of the latter, hypothetically, were extraterrestrial vehicles—which possibility is a long shot—we wonder why an alien civilization advanced enough to send spaceships to the earth would simply flirt with us, showing themselves to only a select few people here and there, and not revealing their purpose. Sightings of alleged flying saucers go back to biblical times. Why would it take so long for them to decide whether or not to establish formal contact? If they thought us rather uninteresting, they would have gone away long ago. Talk of UFOs being on exploratory missions with a policy of "noninterference" sounds too much like *Star Trek.*

Q *Some people who professed to have had encounters with UFOs have passed lie detector tests. Isn't that proof that UFOs are for real?*

A We have never said that UFOs do not exist. We like to keep an open mind because people do see things in the sky that have optical reality.

The lie detector, or polygraph, is another matter. The polygraph records answers to questions as emotional responses, and an expert is required to question the subject and interpret the polygraph tracings. If a person sincerely believed that he or she saw a UFO, the polygraph would indicate that the witness was responding honestly. But the "UFO" may nonetheless have been a weather balloon.

Q *Is it true that mysterious undersea lights seen by ships at sea might have been those of submerged UFOs?*

A We agree that there have been many reported cases of undersea lights seen by sailors. An exact cause or explanation for each of these cases perhaps cannot be given. However, we would point out that there are abundant marine animals and plants that are luminous, sometimes vividly so, from the descriptions we've seen. Some protozoa give off light from minute particles in the protoplasm such that the "sea water seemed on fire." Or sea pens radiate a "golden green light of a wonderful softness." These creatures also can form large underwater "wheels," rotating slowly clockwise or counterclockwise. We suspect that at least some "submerged UFOs" were in fact groups of these luminous creatures.

Q *There are two mysterious places on the earth where the compass points to true north rather than magnetic north. These areas have greater than average UFO activity. Is there any connection?*

A There seems to be a misunderstanding here. There are indeed two areas where the compass points to true north. But it points also to magnetic north because the two are in line with the compass needle. There is nothing mysterious about this. The compass is behaving as it should. These two areas have gained attention because one is the Bermuda Triangle and the other is the Devil's Sea in the Pacific, places of alleged disappearances of planes and ships.

We have no information that would affirm a larger number of UFO sightings in these areas than in any other part of the world.

Q *Isn't there a map of the earth that is thousands of years old yet so accurate that it would have to have been a photograph taken from orbit? This would prove that aliens visited the earth long ago.*

A You are referring to the Peri Re'is map and it is not thousands of years old nor is it that accurate. Peri Re'is was a Turkish admiral and a former pirate who put together a map showing South America, the Atlantic Ocean, and Africa. This was during the 16th century, and for that era, it isn't a bad map. However, if the map is scrutinized, we can see that about 900 miles of South American coastline are missing and the map shows two Amazon rivers. It is therefore not proof that aliens have visited the earth. As far as we are aware, there is no evidence that aliens have ever been on the earth. Scientists do not reject the idea that life, even intelligent life and civilizations, may exist elsewhere, but no evidence has been found that this is so.

Q *Are the reported cattle mutilations caused by UFO aliens or witches' cults?*

A Published reports of cattle and sometimes other farmtype animals being found, particularly in our western states, with tongue, eyes, and sex organs removed have appeared at intervals over many years. To lay the blame on aliens is a rather foolish hypothesis. Yes, there are cults of various kinds that use animals or animal parts in some of their rites. It would seem far easier for the cultists to pick up a stray alley cat for their rituals rather than chase some rancher's cattle out in the middle of nowhere on a dark night. What sensationalist writers fail to mention in their stories on mutilation is that the animals involved here have usually been dead from normal causes for several days before the owner discovers them. In the meantime, smaller animal scavengers have probably attacked the carcass, eating the softest parts such as the tongue. This is another case where something fantastic is proposed, or something sinister implied, when there is a perfectly natural explanation.

Q *Is the movie account of the disappearance of five torpedo bombers in the Bermuda Triangle accurate?*

A No. It is highly romanticized and several facts are omitted or invented. It is true the weather was good when the planes took off in the afternoon of December 5, 1945, but the weather deteriorated as the day progressed. The flight leader, Lt. Taylor, radioed that they were lost. At no time did he report an attack by an alien spaceship.

Taylor was convinced the planes had been blown by high winds across the Florida Keys and into the Gulf of Mexico. Therefore, he flew east to reach Fort Lauderdale. In reality, the planes were still over the Bahamas. In flying east, they flew farther and farther out into the Atlantic until they ran out of gas. One of the last messages heard was Taylor telling his fellow pilots that when they got to ten gallons of gas, they'd all ditch together.

The planes would have ditched in the dark amid 30-foot waves. Chances of surviving under these conditions would be pretty slim. The Mariner search plane, which also disappeared, apparently exploded in midair. A fireball was seen in the sky at the same time that the big plane disappeared from the radar screen.

Q *The* Mary Celeste *must be the greatest sea mystery of all. What do you think happened to her crew?*

A The *Mary Celeste* was found sailing crewless near the Azores in December 1872. You are right in saying it might be the greatest sea mystery. The crew was never found. We will never know exactly what happened. But if we sweep aside all of the sensationalist ideas, certain facts emerge that permit a theory of what happened. We know that the single lifeboat was missing as well as the navigational instruments. A line attached to the vessel had broken. The cargo consisted of several hundred barrels of alcohol. If a ship is in trouble and the captain orders it to be abandoned, the first thing he would reach for are the navigational instruments. One theory is that the Captain, Mr. Briggs, believed either that the flammable cargo of alcohol was about to explode or that the ship was about to sink, and ordered everyone into the single lifeboat, taking his instruments with him. Hoping to get back aboard, they kept a line tied from the lifeboat to the ship, but it parted, and they drifted away to perish, while the *Mary Celeste* sailed on alone to its meeting, days later, with the *Dei Gratia*. But who knows?

Q *Why do some people have more luck than others?*

A Like others, we have met individuals who do seem to be extraordinarily lucky in cards, lotteries, or "stumbling into" the best job. Maybe that is why those of us not so blessed with good luck resort to such helpful objects as a rabbit's foot or other charm. Others swear that certain numbers are lucky for them.

We would have to say that chance plays a large role here. A person who wins a coin flip is not really lucky; there was a 50–50 chance of winning or losing and each new flip presents the same odds no matter how many times a person has won. According to some psychologists, the luck we have in life, good or bad, is brought about by our own previous actions and attitudes leading up to what may appear to be a lucky or unlucky event. The person who gets a very good job may have studied how to handle a job interview and taken the time to dress neatly. To those who did not do so, the person was "lucky." We would call it good planning. While life brings happy or tragic events on occasion, which are beyond our power to circumvent, most luck can be controlled by ourselves.

Q *Where did the idea start that a rainbow brings good luck?*

A The association of a rainbow with good luck probably stems from the Bible. After the great flood, God told Noah He would never again send a flood to destroy mankind and He set His mark or sign in the sky to symbolize this in the form of a rainbow. It is perhaps on this basis that Christians think of people going to heaven by climbing the rainbow or more materialistically finding a pot of gold at the end of the rainbow. However, when you examine early non-Christian cultures, you find that the rainbow often was a symbol of evil. A rainbow arced over a house forecast a death in that house; a man who reached the end of the rainbow would find death rather than gold; out of fear, children were brought inside when a rainbow appeared. So it is a mix of belief. To the scientist, it is simply sunlight splitting into the component parts of light when it encounters droplets of water. By the way, moonlight can also produce a rainbow, although not as colorful.

Q *Why is the number seven considered so lucky?*

A The roots of this belief seem to lie in astronomy. In ancient times, seven planets were known, and the moon had phases that lasted seven days. Four of these phases equaled 28 days, which seemed to govern the menstrual cycle, which in turn governed human life. The Bible is abundant with references to seven, such as the days of creation and the seven marches around Jericho to bring the walls down. The number is still one of importance today; seven days in the week, seven wonders of the world, seven deadly sins as well as seven Christian sacraments and many others, including the gambler trying to roll a seven. If your name has seven letters, that is supposed to make you rather special.

Q *Why is a horseshoe thought to bring good luck?*

A For many centuries the horseshoe has been regarded as not only a symbol of good luck but also a protection against witches or other forms of evil. Why? There are perhaps several reasons. The horseshoe is made of iron and forged with fire. Both iron and fire were viewed as magical by ancient peoples. Perhaps that is why the blacksmith was regarded as a special, almost supernatural person. Even the horseshoe shape, resembling the crescent moon, had magical significance.

Another aspect of the answer is the horse itself. While we talk about the dog being man's best friend, through the ages it was the horse that did a lot of our work—tilling the fields, carrying people and their goods, even being an instrument of war as with the cavalry, and in times of famine a source of food. It is therefore no surprise that the horse and his shoes are held in special regard. Consider also that in more recent times in the old West, horse thieves were hanged.

Q *Why is it thought that throwing salt over the left shoulder will avoid bad luck?*

A Salt is a great preservative and the enemy of decay, as was known from ancient times. It was also thought to be allied to sterility, which is not a good thing. So in effect, salt was a mix of good and bad.

We've all heard the expression "rub salt in the wounds." It hurts. Centuries ago, it was administered as punishment to those who spilled and thus wasted salt. The latter were thought to be under the influence of the Devil. It became the practice for those who accidentally spilled salt to throw some with the right (good) hand over the left (evil) shoulder into the face of the Devil, and so avoid misfortune.

Q *How did the superstition arise that seven years' bad luck will result from breaking a mirror?*

A The mirror captures a person's image, and for ancient peoples, that image, whether reflected in a mirror or in a pool, was the soul of the person. A disturbance of the image was tantamount to loss of one's soul—death or disaster.

With such bad luck in the offing, the question arose, "How long would it last?" The belief in ancient times was seven years because that was how long it was thought the body needed to replace itself and remove all of the bad luck visited upon the original body.

Even today, some remote tribes object to anyone attempting to photograph them, fearing that the capturing of their image on film will cause harm to befall them.

Q *Why do some people cross their fingers when they are hoping for something?*

A This might be traced to early Christian times when people made a cross with the forefinger on the forehead to provide protection from the Devil. Crossing the fingers nowadays is thus still a form of invoking protection. There are any number of superstitions associated with the fingers. It is considered rude to point with your finger, and this may stem from the idea that something bad will happen to whatever is pointed at. In primitive societies, as a sign of mourning a finger was amputated after someone died. There are also separate beliefs concerning the fingernails. They should never be cut on

Sunday or the Devil will have you all week; parings should not fall into the hands of a witch, who can make use of them in casting evil spells.

Q *How did the superstition of "knocking on wood" originate?*

A Most superstitions are concerned with either bringing good luck or preventing bad luck. By knocking on wood, we hope to prevent bad luck relating to something we express as a wish, such as "I think we'll win the game" or "I hope to get the job." It probably originated at a time when early people worshiped certain trees and thought there existed tree spirits who protected them. To touch the tree was to hold evil spirits at bay. To carry a piece of wood from a sacred tree was also a good idea. Archaeological evidence shows that the ancient Chaldeans, Egyptians, Persians, and others regarded trees as sacred. This is really not surprising when you think about the good that comes from trees—we eat the fruit, use the wood for building shelters and objects of art, and even cool ourselves in the shade of a tree.

Q *Why is it considered bad luck to walk under a ladder?*

A Ladders were made to allow people to climb and even in ancient times to get closer to the sky was to become more spiritual and closer to sky gods. Archaeologists regard miniature ladders and depictions of ladders in ancient tombs as symbols of the ascent to a higher and better life. Consider that when a ladder is placed against a wall, it forms a triangle with the ground. To walk through this triangle is to destroy the good luck as well as to reject the symbolism of the ladder itself. In Christian times the triangle became a symbol of the Trinity, and the superstition persisted. Another more practical reason for not walking under a ladder is that you might get a bucket of paint dumped on your head.

Q *Why do sailors consider the albatross a symbol of bad luck?*

A Actually, they don't. For some sailors it is bad luck to kill an albatross. But for many sailors this is not so. Many sailors have gone "fishing" for albatross with line and hook, the bait skimming the surface of the water. In the past, sailors made tobacco pouches out of the skin of the bird, and pipe stems out of the hollow leg bones. There are 13 species of albatross of which the wandering albatross is the most widely known. It is a truly remarkable creature with a wing span of more than 11 feet and the ability to soar effortlessly in the wind for hundreds of miles. Studies of banded albatross have shown they can travel more than 300 miles a day. Essentially, they live their lives in the air, being grounded mostly during breeding times. Despite the fact that they are oceangoing birds, when taken aboard ship, they get seasick!

Q *Do we have free will and control of our future, or is everything predestined?*

A This question has been debated by philosophers for many years and we ourselves have no pat answer. The future can be thought of as "fluid" because a very minor factor or series of factors can greatly alter what might happen. For example, a cat runs across a road in front of a car and the driver puts on the brakes, slowing down the car and delaying its arrival at the next intersection by several seconds. If he had reached the intersection on schedule, he would have collided with a truck and possibly have been killed. Of course, it could be argued that a Supreme Being foreordained that the cat should run across the road. We think it is a better idea to act now to determine your future (getting an education, saving money, and so on) rather than waiting around for the future to happen. True, there are things that no act of free will can accomplish—things that may be physically impossible, like being a lineman on a pro football team if you weigh 110 pounds and are blind. Yet we see no compelling reason to assume that everything in life is predetermined and beyond our own control.

Q *Does the daily horoscope in the newspaper have any real validity?*

A Perhaps, if you are a devout believer in astrology. But even the astrologers are amused that people take it so seriously. An accurate horoscope, according to the astrologers, requires information on all planetary positions and not simply the sun sign, which is what the daily horoscope is based upon.

It seems to us that these horoscopes just dispense good advice such as "take time to relax" or "be careful of business investments." However, it seems absurd that, say, a Leo should avoid travel today, or Libras will find romance tonight. This translates to about 375 million people in the world not traveling today or expecting a romantic encounter tonight because the stars and planets are in a certain position.

More scientific research should be done on the claims of astrology. Even if astrology were totally debunked, this doubtless would not change the minds of many people. Astrology has been around for 5000 years and has always had its followers.

Q *Doesn't the remarkable similarity between identical twins separated from birth show that astrology is not bunk?*

A Only one out of every ten pairs of identical twins studied and reported on were totally separated at birth. Many were raised in the same kind of home environment and economic status. So, a vital question is, do persons of the same age and sex share strong similarities regardless of whether or not they are twins? Psychologist W. J. Wyatt and his colleagues checked this out by testing 13 pairs of twins and 25 pairs of unrelated people of the same age and sex. Areas compared included jobs, politics, hobbies, favorite foods, and so on. The twins turned out to be more similar, but the unrelated pairs showed surprises. According to Wyatt, one pair of unrelated women were "both Baptist; volleyball and tennis are their favorite sports; their favorite subjects in school were English and math (and both listed shorthand as their least favorite); both are studying nursing; and both prefer vacations at historical places. Had these similarities been found in a pair of identical twins (who had been reared apart) they might have been used as evidence for astrology. . . ." We conclude the evidence is weak for any correlation between identical twins' similarities and astrological influence.

Q *How did the zodiac signs originate?*

A More than 5000 years ago, the people of ancient Babylon, and probably other places, would look at the sky on clear nights and interpret certain groups of stars, when connected by imaginary lines, to look like some figure or portrait of a person, animal, or thing. The Greeks later divided the heavens into twelve areas, each dominated by one of the constellations that was perceived as some kind of figure, and called this *zodiakos kyklos*, which means "circle of animals." Each of us, then, is born at a time when the sun appears positioned in one of these twelve areas of the heavens and this becomes our "sign." While astrologers claim that such positions determine one's character and personality, science finds little evidence to support this.

Q *Is more crime committed during the full moon than at other times?*

A Throughout history, many people have believed that the full moon exerted a strong influence on the human mind and emotions, even causing physical changes (such as the werewolf legend). Terms such as "lunatic" and "lunacy" are derived from the Latin word "*luna*," meaning moon. We participated in two separate investigations of police records to check this out. Both studies were conducted over separate six-month periods in Buffalo, New York. At the outset, even policemen we talked to believed that more rapes, murders, and other emotionally violent crimes occurred during the full moon. However, our results showed no greater incidence of these crimes during the full moon than at other times. It would be interesting to compare similar data for other cities.

Q *There are some people whom, meeting them for the first time, I like or dislike immediately. Is there any truth* *to the idea that this is because our auras are in harmony or conflict?*

A The aura has been around a long time. It is the halo depicted around the heads of saints. Modern proponents claim it is a kind of force field of vibrations that exists around everyone. These vibrations reflect the character and personality of an individual. Science has no evidence that such "force fields" exist, but that doesn't mean they don't. Each person, after all, is a bioenergy system that may radiate some force in the same way each of us radiates heat. However, is it not possible that certain subtle external signals such as gesture, voice, intonation, or other physical cues are the real signal to trigger like and dislike rather than some kind of hypothetical aura?

Q *Is the appearance of a comet in the sky considered a good or bad omen?*

A Through the ages, it has been regarded as a bad omen, foreshadowing a famine or plague or disastrous war, even the end of the earth itself. The roots of this belief seem to be the idea that the normally orderly and predictable heavens are suddenly disrupted, suggesting a similar disruption on the earth. Eclipses and meteor appearances can thus be placed in the same category. When such celestial events were indeed followed by some calamity, the notion of a cause-and-effect relationship was of course strengthened, although science would have claimed that it was coincidence. One's own perception is very important here. Julius Caesar was born at the time a comet appeared, and died at a similar time. Depending on what you thought of Caesar, the comet was either good or bad. We don't know of anyone who thinks Mark Twain brought misfortune on the human race; quite the contrary. Yet his birth and death occurred at two successive appearances of Halley's comet.

Q *Can some people have genuine premonitions of future events?*

A There are those who claim to have this power and also ordinary people who maintain they have had such flashes of something that will happen in the near future. Usually the premonition is of something evil or unlucky. There is no scientific evidence that such ability can be exercised at will or even that it occurs at all. Consider that the more information you have bearing upon a future event, such as a presidential election, the better are the odds you will guess what will happen. If a person has a "premonition" that a close relative will die, that person may already have information about the relative's state of health and other circumstances that make the "premonition" probable. It should also be noted that only those premonitions that come true or nearly so are remembered and advertised. Premonitions that do not come to pass are forgotten and not recorded.

Q *How can you say that the future is not seen in dreams when so many dreams are uncannily accurate?*

A It is true that there are many seemingly prophetic dreams that psychics are quick to advertise, but we would have to dispute this on statistical grounds. Realize that all people sleep. Millions of people sleeping mean millions of dreams. If only one person in three had a dream they remembered each day, that would translate to 24 billion recallable dreams each year in the United

States alone. Is it not likely that somewhere in this legion of dreams some of them will hit upon a future event with what seems like, to modify your phrase, uncanny accuracy? At the same time, we would point out that in scientific investigations, there is as yet no hard evidence for or against the idea that some dreams may be prophetic by virtue of some scientific law not yet recognized.

Q *How many dreams does a person have each night and how long do they last?*

A Dreams seem to take place during REM (rapid eye movement) periods. This is a light stage of sleep in which the eyes move and dart around under the eyelids. There are perhaps four or five REM periods a night, which alternate with deeper stages of sleep. During a REM period the sleeper may have one or a series of dreams, some of which last only a few seconds while others are longer. The dreams we remember probably occur during the last REM period of the night. In experiments, subjects deprived of REM sleep become extremely disturbed physically and mentally irrational with hallucinations and blackouts. When allowed to sleep without interruption, these subjects experienced an unusual degree of REM-type sleep. We might conclude from this that dreaming is important to maintaining good health and is not useless at all.

Q *If, in a dream of falling off a cliff, I do not wake before hitting bottom, could I be injured or killed by the fall?*

A No, unless you had in fact gone to sleep on a cliff, had the same dream, and then fell. We have interviewed people who have hit bottom while falling from a cliff in a dream and they are still healthy. It is true that a particularly disturbing dream may cause restlessness during sleep, but it is not lethal.

Q *What is the idea behind crystal gazing?*

A The crystal ball and the rabbit materializing from a top hat are perhaps two of the most recognizable stage props in the whole field of illusion. Historically, the crystal ball has taken a variety of simpler forms. Civilizations as far back as the Assyrians and later the Greeks and Romans gazed into crystals, drops of water, polished metal, or gems. The idea was that one would see there reflected images or appearances that could be interpreted as mystic or religious experiences—perhaps prophecies of good or bad things to come.

The function of the crystal seems to be to concentrate the gaze, but of course science does not attribute any importance to the crystal itself as the source of any visions. If images are seen, they are more likely the product of suggestion and that rich source of strange images, the human subconscious.

Q *Did the prophecies of Nostradamus always turn out correct?*

A There is considerable debate about this because Nostradamus's prophecies, made 400 years ago, were often vague and ambiguous. He wrote them as four-line stanzas, with a deliberate attempt to obscure their meaning. In many cases, the meaning was clear to some people after the event apparently took place. Not all students of Nostradamus agreed on a specific interpretation. For example, the statement, "The armies fought in the air for a long time," might suggest dogfights between enemy aircraft, or could simply mean a battle fought on a mountain. Likewise, the statement, "Across the sky, a long running spark," could mean a meteor or a ballistic missile. Still, many of the French astrologer's predictions, especially for France, seem uncanny.

There are several books about Nostradamus. The interested reader might want to obtain one and make his own interpretations as to whether or not Nostradamus had the power to see into the future.

Q *Was Mother Shipton as great a seer as Nostradamus?*

A Mother Shipton, born in England in 1488, made several predictions, including the deaths of Cardinal Wolsey and Thomas Cromwell, Drake's defeat of the Spanish Armada, and the Great Fire of London. She supposedly wrote several prophetic poems that did not come to light until years after her death. Here is a small example:

> Carriages without horses shall go
> And accidents fill the world with woe.
> Around the world thoughts shall fly,
> In the twinkling of an eye. . . .

This sounds like a forecast of automobiles and radio, but indeed it is a forgery by Charles Hindley, an English editor. Perhaps most of Shipton's "prophecies" were written by someone else after the event. Many of the quatrains of Nostradamus are equally suspect. Nostradamus is probably better known than Mother Shipton, but whether either could be called great is difficult to say because solid scientific proof that anyone can see into the far future is lacking.

Q *Is there any scientific basis for being able to tell the future by reading a person's palms?*

A People have been trying to do this for centuries. We know that palmistry was practiced more than 5000 years ago and probably originated in either China or India. Modern science rejects the notion that the shape of the hand, its various lines and mounts have any bearing upon one's future. Nonetheless, people still spend money to have their palms read. A sharply observant fortune-teller can make almost uncanny statements about a person based upon the handshake and the physical condition of the hand. A hearty handshake indicates vigor and health, a limp handshake, the opposite; a callused hand might belong to a manual laborer, a soft-skinned palm, to an office worker; nails bitten down might suggest chronic worry. Palmists will also take into account other observables such as the clothes worn to make statements about the finances of the client.

Q *A palmist looked at my hand and said I would have three children. Should I believe this?*

A If you did have three children, you might conclude that the palmist had some magical ability or that your hand does reveal your destiny. Yet scientifically, such a conclusion is not necessarily correct. There is a certain logic in the idea that the hands indicate something about a person. For example, callused hands suggest someone involved perhaps in hard physical labor; a soft, pliant hand, a white-collar or studious type; one who gestures frequently, a nervous disposition; and so on. At the other extreme is the notion that specific lines on the hand reveal life expectancy, number of children, love affairs, and other aspects of life. Science does not support this belief, or others that attempt to predict the future. One branch—known as onychomancy—of this "science" studies specifically the fingernails. We have doubts about this also, except that in the case of very dirty fingernails, something of one's character is revealed.

Q *Is it true that Hitler was a great believer in astrology?*

A Hitler was a believer in destiny and did consult with astrologers. To what extent this may have affected history is not exactly known. We know that Goebbels, Hitler's propaganda minister, dumped leaflets from planes over Allied territory with prophecies of the 16th century astrologer Nostradamus that Germany would win the war. It was a good ploy, but it did not work because Nostradamus's prophecies could be interpreted in a number of ways. It is still believed in some quarters that Hitler stopped his Panzer divisions in France when they could easily have destroyed British troops at Dunkirk because an astrologer warned him to stop. We will never know for certain. It now appears from all the evidence that the so-called Hitler diaries were a fake. After Hitler wrote *Mein Kampf*, he didn't do much writing except by dictation.

Q *Is it true that Mark Twain, born in a year which Halley's comet appeared, predicted his own death would occur on the occasion of its next appearance?*

A Absolutely. Halley's comet returns to our solar system every 76 years and appeared in 1910 when Mark Twain died. It was next seen in the sky during March and April of 1986.

Mark Twain was of a psychical bent, and he claimed to have had a dream predicting his brother's death. It was Mark Twain who coined the term "mental telepathy." That Twain did in fact "foretell" his own death with the reappearance of Halley's comet in 1910 is only speculation. Psychiatrists might infer that Twain believed it, and more or less made an autosuggestion that he would die at that time. It seems to us that there is nothing unusual about a person dying at age 76.

Q *Concerning the biblical prophecy that the end of the world will come at the battle of Armageddon, is there such a place, and will the end of the world come in the near future?*

A We need to distinguish here between the destruction of our planet and the end of the human race. In the latter case, we would assume that the earth would be habitable for whatever life survives mankind, perhaps ants. Life on our planet depends for its survival upon the continued functioning of the sun. The sun has enough "oomph" left for many millions of years. Whether we humans behave ourselves sufficiently to take advantage of this planetary longevity is another matter.

Yes, there is a valley of Armageddon in the Middle East. If we are talking about the end of the human race and not the planet, science has no input on this point. Given the explosive situation in the world today, who knows? Realize, however, that groups of people have gathered on hilltops periodically over many decades to witness the destruction of the earth, and it hasn't happened yet.

Q *According to a numerologist, my name is a 2. What does this show about my character and prospects?*

A Probably nothing. The pseudoscience of numerology has been around since the ancient Greeks, Romans, and Egyptians, and why not? Mathematics is the queen of the sciences and aids us in measurement of things and as symbolism for many ideas. However, some people believe that a number has a vibratory influence on individuals and that words can be reduced to a number. Naturally, each number has a certain meaning. For example, your number 2 indicates contrasts and extremes. The number 7 represents the occult and mysterious.

There are several complicated systems, but a common one is to equate the alphabet numbers as in A = 1, B = 2, C = 3, and so on. When I = 9 is reached, start over with J = 1, K = 2, and so on. If a person's name is Joe Smith, then we add 1 + 6 + 5 + 1 + 4 + 9 + 2 + 8 = 36; 3 + 6 = 9 and that is his name number. You can figure out your own the same way. The flaw in the notion that this number has great meaning is that it can change. A different number is perhaps found if we use Joseph Smith, Joseph T. Smith, Joey T. Smith, and so on. While it is an amusing parlor game, there is no scientific foundation on which its principles are based.

Q *Where did the Gypsies get the reputation for having special abilities in fortune-telling and other occult matters?*

A We have no firm answer to this question but we have an opinion. Well known as wanderers, Gypsies have been considered by the local populace, through the centuries, with suspicion, mistrust, and even fear. Perhaps the Gypsies took advantage of this and the gullibility of the curious to ascribe to themselves certain abilities with respect to things supernatural. After all, being nomads, the gypsies had a hard time making a living, and the telling of fortunes and selling of charms or amulets became just another way of making a buck. The origins of the Gypsies are imperfectly known because they have little in the way of written tradition. Their language, however, suggests that they migrated out of northern India, perhaps about the 14th century. Today, they are found all over the world but their numbers are dwindling. Gypsies are true to their own codes of conduct, perhaps no better or worse than those of other societies. It might be noted that they have always suffered severe persecution. During World War II, Hitler wiped out thousands of Gypsies with the same zeal he used against the Jews. Some estimate that Hitler killed 10% of the world population of Gypsies, mostly from eastern Europe. That they continue to endure is to their credit.

Q *Can fortunes be told using dominoes?*

A According to those who believe this, yes. The procedure is to place all the dominoes face down, mix them, and then select three, one at a time. If the same one is drawn twice, its meaning is reinforced, and you are allowed to draw a fourth. In general, the drawing of a blank is bad, especially the double blank. Other doubles are considered good luck in various situations of love, marriage, jobs, and money. A double six is excellent for speculation in money matters such as the stock market. There is no scientific basis for any of this, nor for tea leaves or playing cards for that matter. Nonetheless, it is an amusing pastime if you don't take it too seriously.

Q *Do tarot cards have any occult meaning?*

A Our modern playing cards have probably been derived from them, but the tarot deck was much larger, consisting of 78 cards with a variety of symbols covering all aspects of life—justice, death, fortune, and so on. Tarot cards may have originated as early as the 12th century and were presumably used for fortune-telling, but it is possible that the real origin of the tarot deck was as a tool or means of communication between tribes that could not speak each other's language. Thus, a deck of such cards would be very useful, as symbols, to interchange ideas. For telling fortunes, the deck would be laid out and the interrelationships of the cards permitted some interpretation—good, bad, or indifferent—that would be of interest, depending on the skill and imagination of the fortune-teller. There is no scientific evidence that tarot cards have ever predicted the future for anyone, and they should be regarded merely as entertainment.

Q *Now that 1984 has passed, what can be said of George Orwell as a prophet?*

A We are not sure that it was Orwell's intent to predict what the technology of the 1980s would be like from his perspective of the 1940s. If so, he didn't do very well. The book *1984* makes little or no mention of computers, robots, arms races, or energy crises. The importance of oil is not addressed. Rather than technology, Orwell, who was an ardent anticommunist, wanted to stress the evils of big government in restricting individual freedom. Oddly enough, though, there are smaller, less technically advanced countries around the world whose people enjoy less freedom than in many of the bigger, more advanced countries. We have heard that the original title of Orwell's book was *1948*. Because of constant editorial changes that had to be made (the book came out in 1949), they decided to place the setting farther into the future and therefore switched the last two numbers.

Q *By using a biofeedback machine to increase my alpha waves, will I reduce stress and achieve a higher psychic state?*

A We doubt it. The brain produces several types of waves, including alpha and beta waves. Generally, alpha waves are noted when a person is in a state of relaxation or dreamy reverie. One has to be alert to cause and effect. Do the alpha waves generate relaxation or does relaxation produce alpha waves? Tests show that while one is attentive and has his eyes open, beta waves are produced. The simple act of closing the eyes generates alpha waves. There is no scientific proof that alpha waves create a higher "psychic" state. Science would say that extravagant claims of purging bad habits and curing disease through the use of biofeedback of alpha waves is charlatanism.

Q *Are some people able to make plants grow better simply by talking to them?*

A Some who classify themselves as "psychics" claim they can do this but we have yet to see any sound scientific evidence that this is true. Reactions of plants obtained by hooking them up to lie detectors or other apparatus have failed to be repeatable when studied by other investigators. There is little doubt that some plant growers—those we say have a green thumb—are more successful with their plants than others. Perhaps this is simply an inborn talent in the same way that certain people are more mechanically inclined than others. We know of no experimental method to determine whether carrots scream when they are uprooted, or that the same vegetable doesn't really mind being eaten. Yet some people will believe these claims.

Q *Can plants grow faster and better if you concentrate your mind on them?*

A We have seen no solid evidence to support this claim. There are people who are said to have a green thumb and produce superior flowers and vegetables. To the extent that the green thumbers spend more time, energy, and thought on their plants, you may be right, but to ascribe these results to the pure power of the mind is unwarranted. This is a variation of the ESP idea of psychokinesis, the power of the mind to influence physical objects and move them or even bend spoons. There is no scientific foundation for this. However, don't take our word for it. Experiment yourself. Take two equal-size plants of the same species, give them identical amounts of water, sunlight, and nutrients, but ignore one and focus lovingly on the other. Then observe the effect on their growth. Maybe you will discover something we didn't know before. Otherwise you may find that you can rely on your own observations and experience rather than the word of others.

Q *Do plants have feelings?*

A The notion that plants have feelings come from experiments by Cleve Backster, who hooked up lie detectors to various plants. He claimed that plants responded in a variety of "emotional" circumstances. Plants can show a reaction to such things as rapid changes in temperature, but many of the results of Backster's experiments have not been reproduced in studies by other investigators. This is still an area of the unknown that merits further probing.

Q *What is Kirlian photography?*

A One of the allegedly new parascientific phenomena is a process of recording an image on film in the form of a "glow" that emanates from an object when an electrical field is applied to it.

In Kirlian experimentation, one of the reported sensational cases involves cutting away a portion of a leaf, after which an image of the leaf is recorded on film, including the missing portion of the leaf! It is said that the image of the snipped leaf can be recorded only in the spring, with the implication that by Kirlian photography we can study mysterious life processes. Similar experiments on humans reveal an aura or "flare" from a fingertip, for example.

Introducing an electrical charge into tissue or inanimate objects and recording the discharge on a photographic plate is not new. The technique was named for Russian scientists Semyon and Valentina Kirlian who started their experiments as early as 1939. But although the method is old, there is much new debate about what

Kirlian images represent and what medical or parapsychological aspects they may have.

Q *Some dowsers claim psychic power helps them find water. Couldn't primitive tribes have used such power?*

A We seriously doubt that psychic power is necessary to find water. Subsurface water is more abundant than surface water. You can drill almost anywhere and find water if you go deep enough, even in the Sahara desert. Accordingly, success that is ascribed to water "witching" is undeserved. As for primitive tribes, do not sell them short on intelligence or an understanding of the workings of nature. Even those peoples living in semi-arid or arid areas where water is in short supply would not have needed so-called psychic power to find water. They knew that low areas, especially where vegetation grew better than average, were promising places where water might be near the surface. They also followed animals who seemed to smell or sense water better than humans did and shared the water with the animals. Many desert plants, such as cacti, take up appreciable amounts of water, enough for a pretty good drink, and these plants were known to primitive peoples.

Q *Do some people have a sixth sense of direction?*

A We all know how easy it is to get lost in the bush, even with a compass, or in a strange town. If it seems that others have a "homing instinct," telling them which way to turn, it's not a sixth sense, experts say, but rather an aptitude for observation. Nomadic people or native pathfinders use their senses of sight, sound, and smell to discern direction. Add to this the mental map they construct. A similar "local reference" map is used by cab drivers in New York when they refer to "uptown," "crosstown," and so on. Even the blind have no sixth sense of direction, but rather highly developed natural senses.

Q *If a person can go long distances by astral projection, what happens if they can't get back to their body?*

A This is a highly theoretical question because science has no hard evidence that astral projection occurs. However, in many cultures it is believed that one can leave the body and travel through walls and doors and see events far away while the body remains in a sleeplike state. According to believers, the astral body is still attached to the body by a very elastic silver cord. It is believed that the astral self will be able to return to the body unless the silver cord is severed. The latter event results in death of the physical body and the soul is free to move to a higher plane. While some claim to be able to have out-of-body experiences routinely, we have seen little documented evidence to support it. It should be noted that in certain states of altered consciousness, a feeling of being apart from the physical body may occur, such as under the influence of drugs. This is another fringe area of the unknown that requires rigorous investigation.

Q *Sometimes I feel like I have spoken or done exactly the same thing at a previous time. A friend told me this is called déjà vu. What causes this?*

A It seems to be a common experience. Many of these episodes can probably be explained as scenes or situations viewed during an earlier period of one's life in travel books, films, or other media, and then forgotten. Encountering the same or a similar situation years later could bring on the rush of memory known as déjà vu (French for "already seen"). However, those who believe in reincarnation tend to accept such instances as evidence that the person concerned was here in a previous life. A well known example is the Bridey Murphy case, in which Ruth Simmons, an American housewife, was put under hypnosis in 1952 and detailed a life as a young Irish girl 50 years earlier. Simmons had never been to Ireland. Much of her testimony has been disputed, but we think such cases should be investigated and not simply scoffed at.

Q *ESP seems to apply even in archaeological excavations. What is the story here?*

A The idea of psychic or supernormal power has indeed been applied to archaeological pursuits, although it is not considered a standard or practical technique in normal archaeological science. Psychic archaeology can be compared with map dowsing and water witching as human-potential shortcuts in the age-old quest for hidden treasures of one kind or another. Psychic experience in archaeology cannot yet offer positive proofs of its value and is therefore not yet scientifically respectable.

We usually read about someone with a strange ability to sense the presence of archaeological sites deep in the earth, or by holding an ancient artifact, perhaps an arrowhead, in the hand, a person with psychic talents is said to be able to "see" the maker of the artifact, even across thousands of years.

Q *Is there any validity to the idea that bumps on the head have anything to do with the mind and personality?*

A There is no scientific basis that any particular bump or shape of the head has anything to do with intelligence or personality traits. Such ideas were in vogue early in the 19th century. It was claimed that certain parts of the head reflected an internal organ in the brain that represented a specific trait. Some believed there were 33 such "organs." However, maybe we shouldn't laugh, given the state of knowledge in this field at the time. Today, scientists are subjecting various parts of the brain to electrical stimulation, trying to find specific areas of the brain that control different mental and physical functions. Perhaps the development of phrenology (the study of the conformation of the head) paralleled the rise of chemistry from alchemy. Fields of inquiry must start somewhere.

Q *Can faith healers really cure disease?*

A There are those who swear that ailments they had for years were healed spontaneously and we can't argue with that. Yet we are not sure that faith healing necessarily transcends natural law in that it is miraculous. Despite impressive cures, faith healers have yet to sprout new arms and legs on amputees. It is probable that many sicknesses were psychosomatic, that is, originated in the mind. This is not to say that such illnesses were not real—they were. If a faith healer applies sufficient suggestion to the patient, who in turn sincerely believes in what is being done, this can effect changes in the mind, which has strong control over the physiological processes of the body. Medical science is starting to look into the techniques of the faith healer. How permanent are such cures? We don't know. Records kept are rather spotty.

Q *Are there people who can "see" with their fingers?*

A Like many strange mental phenomena, alleged eyeless vision or DOP (dermo-optical perception) has been reported for many years. The recent revival of interest in it came from the case of a Russian girl who, it was claimed, could read print simply by moving her fingertips over the lines.

Other cases of eyeless vision are stage acts performed by professional mentalists who claim special mental powers. Along with reading print, some who profess to have DOP are said to detect colors with their fingers, or sometimes even with their toes, elbows, or shoulders.

Although entertainers and others who practice eyeless vision must be blindfolded, investigators who are well educated in the art of deception know that blindfolds permit a tiny opening on each side of the nose. They suggest that the only true test of DOP is a metal box fitted over the head and under the chin to eliminate all vision.

Q *What is the explanation for some persons going into a kind of trance and speaking a foreign language they do not know in normal life?*

A This is called "speaking in tongues." We have no firm answer to this question because we have not investigated every such claim. In many of these instances, a person, either physically or mentally ill, will speak nothing but gibberish, but this is interpreted as an "unknown" or ancient language. Intriguing as this sounds, the so-called unknown language is never identified. We are reminded of an uneducated woman who suddenly started to speak Latin. It turned out that she had scrubbed floors and performed other menial tasks for several years in a monastery within earshot of clerics reciting Latin daily. In her feverish state, it is not surprising that she uttered a few incoherent words in Latin, but she most certainly did not recite the orations of Cicero in perfect Latin. We believe most such cases of speaking in tongues can be traced to a natural explanation and not a supernatural one.

Q *There are a group of people—the Holy Ghost people of Appalachia—who handle poisonous snakes and drink strychnine. Why do they do this?*

A To demonstrate faith in the power of the Lord to protect them. In a region of high unemployment, poverty, and distance from business and industrial centers, the importance of religion and its social value is enhanced. Many groups meet three or four times a week for several hours at a time. At these meetings there is much praying and dancing with rhythmic percussion. This leads to a trancelike condition during which poisonous snakes are brought out and handled. Strychnine or other poisons are consumed, usually in small amounts. The snakes themselves are also in a torpor because of the percussion and are less dangerous; if bitten, most people in good health, whether in a religious trance or not, will recover. In small amounts, the body can tolerate strychnine, and indeed it has been used as an antidote for some cases of depressant drug poisoning because it is a stimulant to the central nervous system, causing enhanced sensitivity to sight, sound, and touch. However, we do not recommend it, whatever your religious beliefs.

Index